FUNDAMENTALS OF BIOSTATISTICS

FUNDAMENTALS OF BIOSTATISTICS

Edwin Vann
Wayne State University

D. C. HEATH AND COMPANY
Lexington, Massachusetts Toronto London

This book is dedicated to
those men whom I wish most to emulate:
my Father, Karl Sax, R. C. King, and Bruce Wallace.

Published simultaneously in Canada.

Printed in the United States of America.

International Standard Book Number: 0-669-61747-4

Library of Congress Catalog Card Number: 79-172779

PREFACE, WITH NOTES OF CAUTION TO THE READER

The aim of this book is to present the most basic designs and applications of biostatistical technique to the biologist, who need not have had any previous exposure to this discipline or a mathematical background that includes anything more than a working knowledge of college algebra. For this reason no attempt has been made to present statistical concepts in any but the most rudimentary fashion; consequently, verbal descriptions will be found to accompany most mathematical ones.

This text is based on a lecture course in introductory biometrics that I have presented over a period of several years. I believe that the majority of the statistical problems likely to be encountered by a biologist are represented here in some form. Even if the answer to a particular problem is not covered to the extent needed by an individual and outside help must be sought, a mastery of the procedures described here can likely go a long way in helping to reject the less fruitful paths.

The treatment of biological data is but one side of the coin; the other is the design of the experiment producing the data. Let there be no mistake: An experiment that is well planned—if only in the sense of the data being readily analyzable at the experiment's termination—is a tremendous asset. While it may not solve all the technical problems within the experiment, it will undoubtedly go a long way in minimizing some of them and in providing rapid, straightforward answers to the questions originally posed.

For most of the tests to be later described, there is implied a certain degree of confidence; that is, the experimenter may draw some conclusions as to the relevance of differences or similarities among the various groups being analyzed with a certain likelihood of drawing the correct conclusion. However, these same statistical tests which are designed to help establish the truth can, in a way, be misleading if one blindly accepts or rejects hypotheses about the data solely on the basis of the prescribed level of significance attributed to them by the statistical test. The tests, though objective, are not infallible and, of course, have no minds of their own, so that they cannot determine if the results obtained with them make "biological sense." The experimenter can and must make the ultimate decision as to the acceptability of test results. Thus, the decision must make sense even if it occasionally means the contradiction of the results of the test. A case in point: An ornithologist of my acquaintance showed me some data in which he attempted to quantify the relationship between the vocal patterns of a particular bird species with its activity; for example, nest building, feeding of young, and so forth. However, the data collected in the experiment showing duration of song, interval between songs, ... were so highly variable that it was impossible to show, statistically, any relationship between vocal pattern and behavior. Yet, the data notwithstanding, this ornithologist was himself able to correctly identify the bird's phase of activity with a very high degree of accuracy from tape recordings. Obviously then, the conclusions that would ordinarily be drawn from the results of the test data were meaningless in this instance. Were the tests themselves at fault? No, rather the experiment itself did not measure all of the variables that the ornithologist needed and used in mentally relating vocal pattern with activity.

Another point that must be emphasized deals with the data to be analyzed. The thought of selecting certain observations to the exclusion of others is intuitively unscientific and contrary to the scientists' vow of objectivity. However, oftentimes, one or two observations in a group are conspicuous in that they are vastly larger or smaller than the majority. If they are included in the data to be analyzed, they can occasionally distort the results of the test so that data that would ordinarily be found to support a given hypothesis may be interpreted to refute it. Thus, on occasion, obviously aberrant data must be excluded from the analysis—obviously not without adequate biological justification, however—with the result that the exclusion possibly can help to establish a truth that might otherwise go unnoticed. However, this route is fraught with risk and should be followed only with great caution since, as it sometimes happens, the most unusual (i.e., aberrant) observations are the ones that in the long run lead to the greatest discoveries and understanding of the system under study.

E.V.

ACKNOWLEDGEMENTS

I am indebted to the Literary Executor of the late Sir Ronald A. Fisher, F.R.S., to Dr. Frank Yates, F.R.S., and to Oliver & Boyd, Edinburgh, for permission to reprint Table III from their book *Statistical Tables for Biological, Agricultural, and Medical Research.*

CONTENTS

1

INTRODUCTION TO BASIC PROBABILITIES

Randomness

There are two terms used in statistical analyses with which most everyone has at least a nodding familiarity but to which we will devote some attention because of their importance. They are *random* and *probability*. To begin with an illustration, examine checkerboards (a) and (b) in Figure 1-1. The 50 points in (a) are *uniformly* distributed—

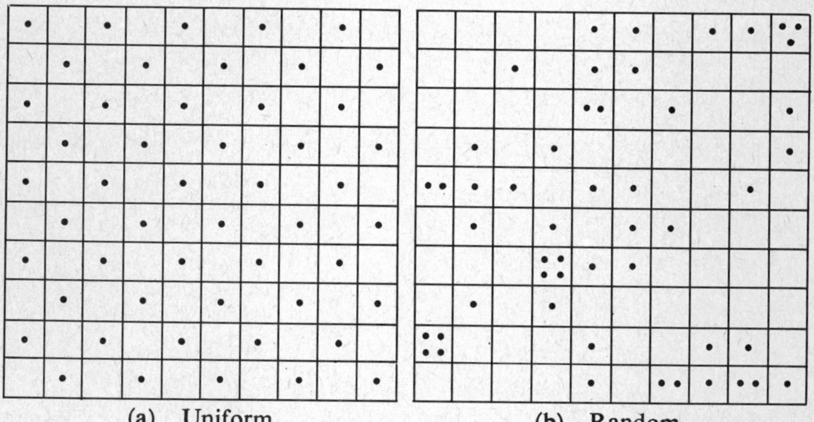

Figure 1-1 A contrast between uniformity and randomness in a distribution of points.

one point in every other square of the 100 squares. The 50 points in checkerboard (b), however, were randomly assigned to the 100 squares on the basis of a random number table. (Such a table can be found in most any handbook of statistical tables and usually consists of at least 10,000 digits, from 0 to 9, which are randomly chosen by machine and entered in the table in their order of occurrence. Since there are ten numbers in the interval 0 to 9, there are generally about 10 percent of each of these numbers in the table.) Each square in checkerboard (b) was assigned a number from 00 to 99. One hundred digits were then extracted from the random number table and arranged in groups of two to produce 50, two-digit numbers. As each two-digit number was recorded it designated a point in the corresponding square. When most persons think of "random," they generally envision something that is uniformly distributed as in checkerboard (a), but as can be seen in (b), randomness denotes complete freedom of choice—within the framework of the system, of course—so that some squares have no points, some have one, some two, and so on, with no predictable pattern.

Probability

This concept can perhaps be best visualized in statistical terms with another example. Assume that you toss a well-worn coin ten times. (The requirement of being well worn so limits the likelihood of its landing on edge that this is nearly nonexistent.) The coin can land, then, in only one of the two other alternative ways: Heads or Tails. One intuitively knows that on any one toss there is a "50 : 50" (i.e., 50%) chance that the coin will land Heads (or will land Tails). Assume that the following results are obtained after 10 tosses: H T H T H H H H H T (7 Heads : 3 Tails). Certainly this deviation from the expected 5 Heads : 5 Tails would cause hardly anyone to lose faith in the fairness of the coin or even to revise his expectations since most everyone realizes that there are many factors (the number of times the coin rotates in mid air, the way it lands, ...) that influence each toss, and with each toss the effects of these factors vary with no set pattern—that is, randomly. Consequently, with all factors acting independently the distribution of Heads and Tails in these 10 tosses was random. Whereas another group of 10 tosses might produce 7 Heads and 3 Tails, 6 Heads and 4 Tails, or even 0 Heads and 10 Tails, the likelihood (that is, the probability) that the coin will land Heads or Tails on any one trial has remained constant: it still is equal to 50%. However, while most samples of ten will not give equal numbers of Heads and Tails, one instinctively expects that in the long run the ratio of Heads to Tails in many trials will approach the ratio 50% Heads : 50% Tails. A probability, then, is nothing more than the limit to which the observed frequency of occurrence tends, when a very large (e.g., infinite) number of observations is made.

Samples and populations

Though we did not label them as such, each group of 10 tosses of the coin constituted a *sample* from a *population* wherein resides every possible value of a variable. For example, a sample of 10 tosses was drawn from the population of all possible coin tosses—clearly an infinitely large number. In the context of the checkerboard example, each sample of 50 randomly distributed points was drawn from the population of all possible random distributions of points. The term "population" can denote a grouping of any size and can be applied to any and all phenomena that have existed in the past, exist at present, and will occur in the future. In biological usage, however, it generally denotes assemblages of organisms that can be as general as "all male *Homo sapiens*" or as specific as "all male *Homo sapiens* of blood type O, born in New York City during the month of July of this year."

A sample has but one use and that is to obtain some insight into the makeup of the population from which it was drawn. Obviously, if this were not the case and each sample were somehow unique, it would be generally useless because of its lack of correspondence to any portion of the universe outside of the sample.

Prediction and inference

Assume that 100 marbles are placed in a bowl; one-fourth (25/100) are Red and three-fourths (75/100) are White for a ratio of 1 Red : 3 White. These marbles will constitute our population. After the marbles are mixed thoroughly, one marble is randomly withdrawn; after its color has been noted and recorded, it is replaced in the bowl and the marbles mixed once more. Due to the replacement of the marble after each draw, it is apparent that the number of draws is infinite. However, before even a single marble has been withdrawn from the bowl one can *predict* (on the basis of the known ratio of Red to White) that on each draw there is a 1/4 probability that the marble will be Red and a 3/4 probability that it will be White. Obviously, one cannot draw a marble that is 3/4 White and 1/4 Red; rather, it is three times more likely that one will draw a White marble than a Red one. It is also to be expected that with larger numbers of trials, or samples, the better will be the likelihood of approaching the true ratio of Red and White marbles since on each draw the marble is randomly chosen and eventually a limit (25% Red and 75% White) will be approached. Even if the proportions of the marbles were not known at the start, after many trials you might have *inferred* that it was "around" 1 : 3 on the basis of the sample results. Therefore, when the probability of an event's occurrence is known before the start of the sampling—which is generally unlikely—one can predict the pattern that the sampling results will follow. Or, if one must rely on only the results of the sampling data—which is by far the most typical situation—one can infer the expected proportion or probability of the event's occurrence on any trial. It should be emphasized, however, that even though larger numbers of samples or, taken another way, larger numbers of observations per sample are made that increase the chance of obtaining the expected long-run results or of inferring the true probability of an event's occurrence, there is no guarantee that you will ever be completely accurate. Indeed, in only an infinitesimally small number of cases would this be true. However, one must always adhere to the doctrine: *The larger the sample, the more confidence can be placed in the reliability of the conclusions that are drawn regarding the makeup of the population.*

More complex probabilities

We can consider a slightly more complicated problem, still within the framework of the coin tossing game. For example, one can ask about the expected outcomes when tossing two coins, which is the same as combining the results of single tosses in groups of two.

Consider two tosses of a coin. There are only two possible outcomes: Each can land either Heads or Tails. With regard to both outcomes, considered as a unit, there are however, four possibilities:

1. Both land Heads.
2. The first lands Heads, the second Tails.
3. The first lands Tails, the second Heads.
4. Both land Tails.

One can visualize this situation by means of another checkerboard, as in Figure 1-2.

First \ Second	*H*	*T*
H	*HH*	*HT*
T	*TH*	*TT*

Figure 1-2

There are $2 \times 2 = 4$ possible outcomes, since for each of the two possible outcomes for the first toss, there are two possible outcomes for the second. Taken another way, imagine two boxes—one for each coin. The first box can be filled in one of two ways, so a "2" is entered; the second box can likewise be filled in one of two ways, so a "2" is also entered.

2		2

The number of ways both boxes can be filled is

$$2 \times 2 = 4$$

Consider now a problem that has some biological relevance. DNA is generally composed of four nucleotides: Adenine (A), Guanine (G), Cytosine (C), and Thymine (T). In chains of 2 nucleotides the first may be either A, G, C, or T and the second also A, G, C, or T. If we set up a checkerboard (Figure 1-3),

First \ Second	A	G	C	T
A	AA	AG	AC	AT
G	GA	GG	GC	GT
C	CA	CG	CC	CT
T	TA	TG	TC	TT

Figure 1-3

it is obvious that there are

$$\boxed{4} \times \boxed{4} = 16$$

possibilities for pairs of these nucleotides.

In a chain of *four* nucleotides there are

$$4 \times 4 \times 4 \times 4 = 4^4 = 256 \text{ combinations}$$

It is now possible to formulate a rule that will describe these events and any others of this type: *If a first event can occur n_1 ways, a second in n_2 ways, . . . , a jth in n_j ways, then the number of possible outcomes is*

$$n_1 \times n_2 \times \cdots \times n_j$$

It is sometimes confusing to note that two apparently similar events have different probabilities if one knows more about one event than the other. For example, as a prediction about paired combinations of four nucleotides, you can consult the checkerboard and see that there is a probability of one in sixteen that the final result will be, say, TT. But, if the first nucleotide is known to be T, the probability is now one in four that the second will provide the final TT result. Although this may seem inconsistent it is not, as the difference lies in the fact that the probabilities are based on different possibilities in the two situations. Before the first nucleotide was known, the likelihood of observing TT was one of sixteen equally likely combinations: AA, AG, . . . , TT. But, when the first nucleotide was fixed at T, the other twelve that did not include T as the first choice were eliminated. As there are only four possible outcomes remaining (TA, TG, TC, TT), the probability of a TT event has increased to one in four.

All of the examples that have been considered depend on the *independence* of the events whose probabilities are being computed. So, even if a chain of 99 T nucleotides is observed—and if there is no a priori reason to assume anything other than a random pattern—one can still assume that the probability of the 100th being T remains equal to one-fourth.

In view of these observations we can modify our first rule as follows: *If the probability of occurrence of a first event is p_1, a second is p_2, . . . , a jth is p_j, then the probability that they will all occur together is*

$$p_1 \times p_2 \times \cdots \times p_j$$

Thus, if the probability of A = 1/4 = p_1, …, T = 1/4 = p_4, the probability of observing any combination of nucleotides in, say, groups of two is as shown in Table 1-1. Note that the total probability of occurrence of all combinations of events equals unity (16 × 1/16); this indicates that all (100 %) of the possible combinations of pairs of nucleotides have been considered. Each of the sixteen combinations has an equal probability of occurrence (1/16) but you will note that the second (AG) and fifth (GA), third (AC) and ninth (CA), fourth (AT) and thirteenth (TA), and so forth, have one thing in common: they are *permutations*—that is, rearrangements—of the same event. Consequently, the occurrence of, say, the combination of one A and one G with the *order not specified* is fulfilled by either the second or the fifth combination.

This introduces us to the last of our most basic rules of probability: *The probability of occurrence of one or more mutually exclusive events is the sum of the separate probabilities.* (The term "mutually exclusive"

Table 1-1 Probabilities of occurrence of all possible combinations of two nucleotides

Combination number	Nucleotide number 1	Nucleotide number 2	Probability
1	A	A	$p_1 \times p_1 = \frac{1}{4} \times \frac{1}{4} = \frac{1}{16}$
2	A	G	$p_1 \times p_2 = \frac{1}{4} \times \frac{1}{4} = \frac{1}{16}$
3	A	C	$p_1 \times p_3 = \frac{1}{4} \times \frac{1}{4} = \frac{1}{16}$
4	A	T	—
5	G	A	—
6	G	G	—
7	G	C	—
8	G	T	—
9	C	A	—
10	C	G	—
11	C	C	—
12	C	T	—
13	T	A	—
14	T	G	—
15	T	C	—
16	T	T	$p_4 \times p_4 = \frac{1}{4} \times \frac{1}{4} = \frac{1}{16}$
			Total = 1.00

means only that if one such event occurs it excludes the occurrence of any other. A pair of nucleotides, for example, cannot be both AG and GA at the same time, as the occurrence of one combination precludes the occurrence of the other.) In our last example the probability of obtaining AG, in that order, equals $1/4 \times 1/4 = 1/16$; the probability of obtaining GA, in that order, is $1/4 \times 1/4 = 1/16$. The probability of obtaining an A and G combination in any order is, then,

$$1/4 \times 1/4 = 1/16 \qquad \text{for} \quad \text{AG}$$

and

$$1/4 \times 1/4 = 1/16 \qquad \text{for} \quad \text{GA}$$

for a final probability of occurrence of $1/16 + 1/16 = 1/8$.

PROBLEMS

1. In whites in the United States the sex ratio (males to females) observed at birth is about 1.06; that is, approximately 106 males for every 100 females. Since this ratio predicts roughly equal numbers of the two sexes, it is customary to compute sex ratio problems letting frequency males = frequency females = 0.50 = 1/2. (a) In a family with 3 children what is the probability that all are males? (b) What is the probability that all are females? (c) What is the probability that *at least* one will be female?
2. Assume that the combination of a new type of lock consists of four letters. (a) What is the total number of different four letter combinations possible? (b) What is the probability of observing a combination with the letters *ZKWK*? (c) What is the probability of a combination with exactly two *L*'s?
3. A standard deck of playing cards contains 52 members. (a) What is the probability that a single card drawn will be either a king *or* a spade? (b) If one card is drawn from the deck but *not* replaced before a second is drawn, what is the probability that the second card will be a king *and* not a spade.
4. A penny, nickel, and dime are each tossed three times. What is the probability that one would obtain *at least* two Heads on the penny, *at least* one Head on the nickel, and no Heads on the dime?

ANSWERS TO PROBLEMS

1. (a) $0.50 \times 0.50 \times 0.50 = 1/2 \times 1/2 \times 1/2 = (1/2)^3 = 1/8$
   (b) Same as part (a)

(c) Any of the following 7 combinations fulfill the requirement of *at least* one female:

F M M = 1/8	F F M = 1/8	F F F = 1/8
M F M = 1/8	F M F = 1/8	
M M F = 1/8	M F F = 1/8	

Third column denotes the probability of all children being female: 1/8. Second and first columns denote all permutations of two females and one female per family, respectively: 3(1/8) and 3(1/8). Total probability of occurrence of any combination is the sum of all the single probabilities: $1/8 + 3/8 + 3/8 = 7/8$.

2. (a) There are twenty-six choices for each of the letters of the combination. Total number possible is $26 \times 26 \times 26 \times 26 = (26)^4 = 456{,}976$

(b) $1/26 \times 1/26 \times 1/26 \times 1/26 = (1/26)^4 = 1/456{,}976$

(c)

L L – – = $(1/26)^4$	L – L – = $(1/26)^4$	L – – L = $(1/26)^4$
– L L – =	– L – L =	
– – L L =		

Probability of any of the acceptable combinations occurring is $6(1/26)^4 = 6/456{,}976$

3. (a) Probability of king: 4/52; probability of spade that is not a king: 12/52 Probability of either occurrence: $4/52 + 12/52 = 16/52$

(b) If the first card is not a king, the probability of the second card being one is 4/51; or, if the first card is a king, the probability becomes 3/51. If the first card is not a spade, the probability of the second being one is 13/51; or, if the first is a spade, the probability becomes 12/51. Probability of all possible occurrences:
$(4/51 + 3/51) \times (13/51 + 12/51) = 7/51 \times 25/51 = 175/2601$

4.

Penny	Nickel		Dime
H H T = 1/8	H T T = 1/8	H H T = 1/8	T T T = 1/8
H T H =	T H T =	H T H =	
T H H =	T T H =	T H H =	
H H H =	H H H =		

Final probability is $4(1/8) \times 7(1/8) \times 1/8 = 28/512$

2

DESCRIPTION OF DATA

Some basic notation

Any measurable characteristic which shows variability (e.g., weight, height, age, ...) is called a *variable*. Customarily symbols are used to denote such quantities. For example, if we follow the usual convention and let the letter X denote some variable (such as height), then the term X_i ("X sub i") can be used to denote a particular measurement of height, although at the moment no one particular observation need be in mind. But, if it becomes desirable to refer to a particular observation, like the third height recorded in a sample of 10, a number may be substituted for the subscript i and the notation becomes $X_3 = 42$ cm. Thus, X_i can denote $X_1, X_2, X_3, \ldots, X_{10}$ in this example. The last observation in any sample is usually denoted X_n, and in this case $n = 10$.

The upper-case Greek letter Σ ("sigma") is used as an abbreviation denoting a sum, so that the expression

$$X_1 + X_2 + \cdots + X_{10}$$

can be rewritten in shortened form as

$$\sum_{i=1}^{n=10} X_i \quad \text{or} \quad \sum_{i}^{n} X_i$$

still meaning that all values of the variable X, from the first to the nth, are to be summed.

Significance of figures

So far we have described quantities which exhibit variation by the rather nebulous term "variable." As this does not fully describe the situation we must begin our discussion by distinguishing between the two principal types of variables: *continuous* and *discrete*.

If, by way of example, one counts 5 colonies of bacteria on a culture plate, it is useless to record this number as, say 5.00, since 5 by itself describes the number of colonies as perfectly and as accurately as is possible. In other words, in this example we are dealing with a discrete variable; there is no way to subdivide the interval between 4 and 5 or 5 and 6 bacteria.

Continuous variables, however, are those which can take on any value in a given interval. To use a different example, assume that one records the weight of an organ of the body to the nearest gram; further assume that the weight is found to be slightly in excess of 5 gm. If the weighing device is of even modest accuracy, the measurement can be refined to the nearest tenth (0.1) gm, thereby implying that the difference between 5 and 6 gm is now divided into ten 1/10 gm intervals. Assuming that the device is extremely accurate one might continue dividing this 1-gm interval until, with an infinitely accurate device, the difference between 5 and 6 gm is divided into an infinity of smaller intervals. As should now be apparent, some sort of convention for describing continuous variables must be followed. The convention is based on two premises: First, the observed value of a variable is assumed to be the midpoint of the range of the measurement. Thus, the value 5 gm must lie in an interval 5 $\pm$ ("plus or minus") some number. Second, since the last digit of the measurement should imply something about its accuracy, the interval about the variable is as long as the smallest division of the measurement; thus 5 gm denotes that the variable was measured to within 1 gm of its true weight or else it would have been recorded either as 4 or 6 gm. Consequently, the true value of the measurement lies somewhere in the 1-gm interval 5 ± 0.5 gm; i.e., 4.50... and 5.49... gm where "..." denotes further subdivisions of the interval and only the *first* digit following the decimal is considered significant. The upper value of the interval about 5 gm must not equal 5.50... for if it did, the true value could be designated as either 5 or 6 gm since the intervals about both of these would overlap at this value. By the addition of a zero in the first (tenths) decimal place (e.g., 5.0 gm) one implies a measurement that is accurate to within 1/10 gm of the true value and so the interval in this case would be 1/2 of 1/10 gm above and 1/2 of 1/10 gm below 5; that is, $5 \pm 1/20$ gm. The results of the refinements of continuous measurements are shown in Table 2-1.

Very often one encounters the problem: How large (or small) should the units of measurement be? To the nearest gram, or tenth or

Table 2-1 Notations describing increasingly accurate measurements of a continuous variable with value 6.0000

Measurement (gm)	Interval around the true value (gm)	Width of the interval (gm)
6	5.50 . . . through 6.49 . . .	1
6.0	5.950 . . . through 6.049 . . .	.1
6.00	5.9950 . . . through 6.0049 . . .	.01
6.000	5.99950 . . . through 6.00049 . . .	.001
6.0000	5.999950 . . . through 6.000049 . . .	.0001

hundredth; to the nearest millimeter, or tenth or hundredth, and so forth? One rule of thumb is: *If the observations are arranged in order of size, the number of units of measurement in the interval between the smallest and the largest observations should be between 30 and 300.* So, if one makes a series of weight measurements to the nearest gram and the largest observation is 5 and the smallest 1 gm, since there are only four divisions in this interval, by application of the 30–300 rule, one should record the weights to one significant decimal place. If this is done and the weights are now recorded as, say, 5.2 and 1.2 gm the number of units separating them is 40, which is within the limits of the rule. By the use of this rule, an error of one measurement unit could cause no error that would exceed a maximum of 1/30 or 3.3%. That is, if a weight of 5 gm is incorrectly recorded as 6, then the recorded value is incorrect by a factor of 6/5 = 1.2. However, 5.1 recorded instead of 5.2 (also a one unit inaccuracy) produces a measurement which is in error only by a factor of 5.2/5.1 = 1.02.

Rules of significance

There are some general principles for the handling of discrete and continuous variables which will lessen the chance of committing errors in describing the accuracy of the observations. Simpson et al.† have neatly combined these principles into several straightforward dicta. They are, in modified form:

† G. G. Simpson, A. Roe, and R. C. Lewontin (1960). *Quantitative Zoology*, revised edition. New York: Harcourt Brace Jovanovich.

1. In any computation involving only discrete numbers, all the resulting digits are significant.

$$\frac{7+9+27}{3}=\frac{43}{3}=14.333\ldots$$

2. When continuous numbers are added or subtracted, the answer has one less significant figure than the number with the fewest significant figures.

$$2.03+3.1151=5.1$$

3. When discrete and continuous numbers are added, the sum has as many significant figures as has the continuous number.

$$2.03+10=12.03$$

4. When a continuous number is multiplied by a discrete number, the product has one less significant figure than has the continuous number.

$$2.03\times 9=18.3$$

5. When a continuous number is divided by a discrete number, the quotient has as many significant figures as has the continuous number.

$$9.623/9=1.069$$

6. When two continuous numbers are multiplied or divided, the answer has one less significant figure than the number with the least significant figures.

$$2.03\times 3.115=6.3$$

7. The square root of a continuous number has the same number of significant figures as does the number itself.

$$\sqrt{9.44}=3.07$$

8. All of these previous operations produce one more broadly significant figure than strictly significant figures.

Rounding

Quite often one obtains a value having more figures than are really significant and so the excess digits have to be "rounded off." Simply dropping the superfluous digit(s) is not satisfactory as this may well introduce an error into the data. For example, let us say that we wish to round off the value calculated for the average density of the earth:

344.7 pounds per cubic foot. Logically, if one wished to round off this number to the nearest whole pound it would not be correct to set it equal to 344 pounds since the value is clearly closer to 345 pounds. So, one can adopt a convention that works quite well: *Round the last significant digit downward if the following (nonsignificant) number is less than 5; round it upward if the following (nonsignificant) number is greater than 5.* Under this rule 344.7 would be rounded to 345. But, what happens when the first nonsignificant number is 5? A procedure that works well in the long run is to round *upward* when the last significant digit is *odd* and *downward* when it is *even*, so that 123.5 becomes 124 while 124.5 remains at 124 after rounding.

Estimators of central tendency

A population contains all possible values of a variable and may be of any size. While it is usually difficult or even impossible to observe every member of a population—as most populations are usually defined—it is possible to draw one or more samples from a population and use these to make an inference about the population's composition. But before one can do this, the observations must be arranged in a fashion such that their particular characteristics are made readily obvious and, thus, more readily available for interpretation. Fortunately, one can take advantage of some of the general physical properties of populations and samples in order to accomplish this end. For example, the vast majority of variables with which a biologist is concerned are distributed in such a way that certain classes of values are more frequent than others, and the number of observations in the various classes usually decreases as one goes further away from the most common (frequent) values. In other words, most biological distributions have a tendency to cluster around a certain class—that is, they show a *central tendency* of the sort illustrated in Figure 2-1. The type of graph shown is called a *frequency distribution* because it describes just how frequently one observes a given number of yeast cells in a hemocytometer square. It is apparent that the observations tend to cluster more toward the lower than the higher numbers of cells per square.

Whenever one refers to an average (more correctly *mean*) value, a specific characteristic of the central tendency of the distribution of variables is implied. The *arithmetic mean* (the one with which most persons are familiar) can be represented by the Greek letter μ ("mu")

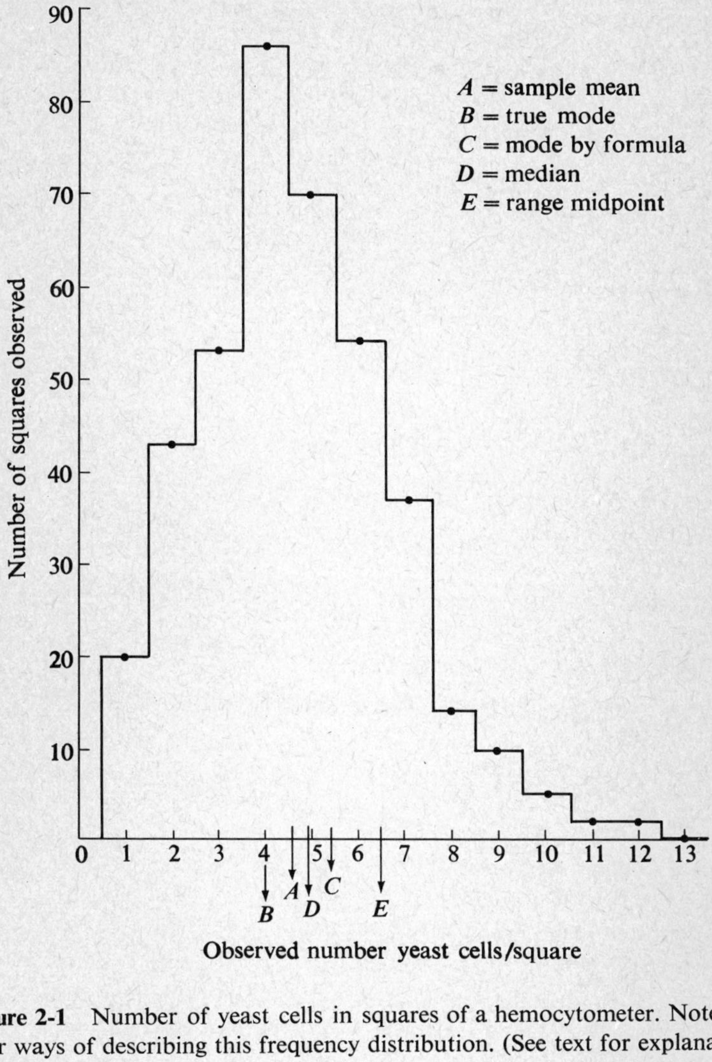

Figure 2-1 Number of yeast cells in squares of a hemocytometer. Note the four ways of describing this frequency distribution. (See text for explanation of terms.)

when it is a *parameter* (characteristic of a population) and, $\overline{X}$ when it is a *statistic* (characteristic of a sample). Since the sample is a representative of the population, it follows that $\overline{X}$ is an estimator of μ. While $\overline{X}$ varies with the particular sample, due to the randomness of the observations comprising the sample, the values in the population are fixed and so μ, too, is a fixed quantity. For example, consider a box containing 100 cards each one numbered consecutively. If the 100 cards are considered to be a population, then the true value of the population mean is

$$\mu = \frac{1 + 2 + \cdots + 100}{100} = \frac{5059}{100} = 50.59$$

But a sample of 10 cards drawn from the box might result in the values: 58, 64, 21, 50, 31, 16, 52, 97, 17, 1 so that this particular estimate of μ is found to be

$$\overline{X} = \frac{58 + 64 + \cdots + 1}{10} = 40.70$$

The formula for the mean can be symbolized as

$$\sum_{i}^{n} X_i/n$$

which can be read as "the mean value of the variable per observation."

The mean is really a focal point in a distribution and as a general rule does not even represent a real observation. Take the case of the array: 6, 7, 8, 9, 92. Here, the mean equals 24.4 yet no such observed value, or one even similar to it, was recorded. The purpose in finding the mean is to measure observations as deviations from it $(X_i - \overline{X})$ so that the total *magnitudes* of the deviations on either side of it (i.e., larger and smaller) are equal. For example, Table 2-2 shows the deviations about the mean in the sample of five numbers noted above. Notice first that the single observation $(X_i = 92)$ that is greater than the mean completely balances out the deviations of the four lesser values, and second, since the mean is a measure of central tendency, the sum of the deviations about it must always be equal to zero.

Another measure of central tendency, though not widely used, is the *geometric mean*. This quantity as applied to biological problems is used primarily to determine rates and ratios in systems whose characteristics change with time. For example, assume that one wishes to

Table 2-2 Deviations about the mean of an array of five numbers

X_i	$X_i - \bar{X}$
6	$6 - 24.4 = -18.4$
7	$7 - 24.4 = -17.4$
8	$8 - 24.4 = -18.4$
9	$9 - 24.4 = -19.4$
92	$92 - 24.4 = 67.6$
$\sum X_i = 122$	
$\sum X_i/n = 24.4 = \bar{X}$	$\sum = 0.0$

determine the mean rate of growth of a colony of bacteria. The arithmetic mean is not applicable because bacteria grow not in an arithmetic but in geometric fashion: $1 \rightarrow 2 \rightarrow 4 \rightarrow 8 \rightarrow 16 \rightarrow \rightarrow$, etc.

$$\text{Geometric mean} = \sqrt[n]{X_1 \cdot X_2 \cdot X_3 \cdots X_n}$$

where X_i is the value of the observation and n is the number of observations (specifically, intervals) used in the calculation. The data of Table 2-3

Table 2-3 Growth of a bacterial culture begun with 10 cells

Time (hr)	Number of cells	Growth ratio relative to origin
0	10	
1	43	43/10 = 4.3
2	167	167/10 = 16.7
3	620	620/10 = 62.0
4	2719	2719/10 = 271.9

show the results of cell counts in an actively growing bacterial culture that was begun with 10 cells (with a characteristic doubling time of approximately 1/2 hour) and was sampled at hourly intervals. In this example the geometric mean of the cell number is

$$\sqrt[4]{(43)(167)(620)(2719)} = \sqrt[4]{12{,}105{,}586{,}180} = 331.7$$

While this is a perfectly valid statistic and measure of central tendency (provided no number in a series of observations equals zero), the mean value in this case, 331.7 cells, has little intuitive meaning, so that given just this number someone not familiar with the nature of the system would probably have a difficult time envisioning it. Another way of applying this statistic to these data may have a greater intuitive appeal. The third column of Table 2-3 shows the proportional change in observed numbers of cells each hour, relative to the size of the culture at time zero. For these values

$$\text{Geometric mean} = \sqrt[4]{(4.3)(16.7)(62.0)(271.9)} = \sqrt[4]{1{,}210{,}558.6}$$
$$= 33.2$$

indicating an average proportional increase of 33.2 times over the 4-hour time period. Still, even this value is not readily visualizable in the context of this system. Probably the most frequently used application of the geometric principle—possibly because of its intuitive appeal—has to do with average percent (rate) of change. In this bacterial culture the original number of founders was 10, but 4 hours later the number was estimated to be 2,719; thus, the net change was 27,190% per 4 hours. If one takes one-fourth of that value (i.e., 6,797%) as the hourly percent increase and computes a 6,797% increase per hour for 4 hours, the number of cells predicted is $>> 2{,}719$ (the $>$ symbol being read as "larger than"). The obvious inaccuracy of this hourly percentage change can readily be seen from an examination of the table, where the growth ratios per hour are about 4 (e.g., $43/10 = 4.3, \ldots, 2719/620 = 4.4$). The correct growth figure, then, should be much closer to 400% than to 6,797% since the numbers of cells are increasing geometrically each generation. The average percent of change per hour can be computed from

$$P_f = P_0(1 + r)^n$$

where P_0 = population size at time zero,
P_f = population size at the end of the time period,
r = unknown, relative rate of change per interval expressed as a decimal, and
n = number of intervals.

For the data of Table 2-3 this equation becomes

$$2{,}719 = 10(1 + r)^4$$

Solving by logarithms

$$\begin{aligned}
3.43441 &= 1.00000 + 4 \log (1 + r) \\
0.60860 &= \log (1 + r) \\
4.061 &= 1 + r \\
3.061 &= r \\
306.1\% \text{ per hour} &= \text{rate of increase}
\end{aligned}$$

which is obviously more in keeping with the estimates made by inspection of the hourly observations alone.

Another measure of central tendency is the *median*, or the value in a sample or population that has equal numbers of observations above and below it. In an odd-numbered array of observations the median is itself an observation. For example,

$$\text{Median } [6, 7, 8, 9, 92] = 8$$

In an even-numbered array, however, the median is a nonobserved number in that no middle observation exists. However, one can determine the median value in such an array by adding the two most central values and dividing by 2:

$$\text{Median } [6, 7, 8, 9] = (7 + 8)/2 = 7.5$$

The median is an extremely useful quantity (especially when used in conjunction with the arithmetic mean to describe observations) since it is less apt than the mean to be distorted by extreme values. For example, as we noted in our earlier discusssion, the arithmetic mean provides a point around which the deviations of the observations to either side of the mean cancel out. Thus, both the arrays

1. $\bar{X}[24.2, 24.4, 24.6] = 24.4$ and
2. $\bar{X}[6, 7, 8, 9, 92] = 24.4$

have the same mean, though the mean of array number 2 is influenced as much by the single value $X_i = 92$ as by the four other observed values of this array, combined. However, the medians for the two arrays (24.4 and 8, respectively) differ markedly, so that even without a knowledge of the observations themselves this statistic presents a strong indication of the differences between the sets of observations.

Frequency distributions for most biological variables tend to exhibit a symmetry (often bell-shaped) and to cluster around a point.

That point is often the mean. However, in extremely *skewed*, or asymmetrical, distributions having a pronounced tail to the right or to the left of the central tendency, the mean is not really a nucleus or point of concentration of values. Reexamine Figure 2-1, where the frequency distribution exhibits a skewness (tail) to the right. Then, for describing distributions in which there is a departure from symmetry—that is, a noticeable amount of skewness—it is necessary to have a measure other than the mean to designate clustering. Logically, it should be the value or class having the greatest number of observations; such a group is called the *modal* class. There are two ways of estimating this quantity. First, the easiest and most straightforward is by inspection: The data are grouped into equal intervals and the class with the greatest frequency is designated the modal class. The second method is usually more accurate but requires a formula of estimate:

$$\text{Mode} = \bar{X} - 3(\bar{X} - \text{Median}) = 3\ \text{Median} - 2\bar{X}$$

This formula should not be used for extremely skewed distributions for in such instances an estimate by inspection is the most accurate method.

Measures of dispersion

So far we have looked at several methods of describing biological distributions by means of their central tendencies. While the sample mean, $\bar{X}$, is generally the most useful all-round statistic, by itself its description of the distribution often leaves much to be desired since it tends to be strongly influenced by even a relatively few extreme values. What is needed, then, are some methods of describing a distribution not by its central tendency but by the opposite: its *degree of dispersion about the central tendency*.

One such measure of dispersion is the *range*: this is simply the difference between the highest and lowest observed values. As the number of dots on the face of a die (plural: dice), for example, can vary from 1 to 6, the range of possible observable values in this population equals $6 - 1 = 5$. In reality, while the range of the sample is readily obtainable, the range of the population is almost never known. As a measure of dispersion it has severe drawbacks, so that for this purpose it is of only limited utility. For example, the observed range is generally completely dependent on the number of observations in the sample, so that simply by increasing the number of observations the observed range increases. Second, unless the frequencies of the various classes are given, the real meaning of the range is obscure since the extreme observed values, though perhaps relatively infrequent, will be given the same weight as the relatively more common middle values. See Figure 2-1 with the aim of noting just how much information can be obtained from the value of the range, 11.

However, in some cases this statistic does have a real utility. For example, in taxonomic classifications where one class (genus, species, sex, etc.) has a *sharply delineated* range of variability for some metric character, two different types of plants or animals can be distinguished by the observed range: by the range in the number of bristles in the case of some insects, in the number of fins in the case of some fish, and so on.

The ideal statistic, then, would be one that would utilize not only the range of variability (dispersion) but also the number of measurements in a certain class (frequency) as well as the mean of the distribution (central tendency). The quantity which is the best overall is the *variance*, or *mean square*, represented by σ^2 (the lower-case Greek letter sigma) for a population and s^2 for a sample. The variance is defined as the sum of the squares of the differences (that is, deviations) between the observations and the mean, divided by $n - 1$ where n is equal to the number of observations in the sample of population. The variance can be defined as

$$\begin{aligned}\sigma^2 &= [(X_1 - \mu)^2 + (X_2 - \mu)^2 + \cdots + (X_n - \mu)^2]/n - 1\\ &= \sum_i^n (X_i - \mu)^2/n - 1\end{aligned}$$

for a population, and

$$\begin{aligned}s^2 &= [(X_1 - \overline{X})^2 + (X_2 - \overline{X})^2 + \cdots + (X_n - \overline{X})]/n - 1\\ &= \sum_i^n (X_i - \overline{X})^2/n - 1\end{aligned}$$

for a sample.

One can note that squaring the difference $(X_i - \overline{X})$ or $(X_i - \mu)$ makes all deviations about the mean effectively positive, so that this quantity gives no consideration to the direction of the deviation; that is, if the observation is greater or less than the mean. As a result, the variance is *approximately* the square of the average deviation about the mean—without regard to the direction of the deviation.

The numerator of the variance is called the *sum of squares*, hereafter abbreviated *SS*. The denominator, $n - 1$, is termed *degrees of freedom* and is abbreviated *df*. In this context degrees of freedom denotes the number of independent deviations measured by the variance. The sum of squares can be converted to a *working formula* which is handled more readily on a desk calculator than the previous definition formula. For example, for a sample

$$SS = \sum_i^n X_i^2 - \left(\sum_i^n X_i\right)^2 \Big/ n$$

Here, each observation is squared and all are added together for the first term while for the second, all observations are added together and the total squared before dividing by n. We can now write

$$s^2 = SS/df$$

The quantity $\sum X_i^2$ is called the "unadjusted" sum of squares and $(\sum X_i)^2/n$ the "correction term," abbreviated *CT*. While this distinction is not of great importance here, in our later discussions it will be.

It is often convenient to use a different form of the variance by extracting the square root. While this procedure produces absolutely no new information on the degree of dispersion among the observations, it does reduce the size of the deviations about the mean (which are originally squared for the calculation of variance) to one more directly relatable to the deviations themselves. This new quantity is called the *standard deviation* and is

$$\sigma = \sqrt{\sigma^2}$$

for a population and

$$s = \sqrt{s^2}$$

for a sample.

Thus far we have been concerned with the numerous ways of describing samples drawn from populations. But it follows that while $\overline{X}$ is an estimator of μ and s^2 of σ^2, it is unlikely that any single value of $\overline{X}$ or s^2 will precisely equal μ or σ^2. One intuitively expects, though, a group of several sample means to show less variability, or dispersion around the population mean μ, than an equal number of individual observations drawn from the population. Indeed, if this were not the case there would be no reason to ever make more than one observation from any population since any single observation would provide just as good an estimate of μ. Then, just as one can plot a frequency distribution for observations in a sample, one can also plot a *frequency distribution of the means* of several samples, each $\bar{X}_i$ being treated as a single observation, and, as such, one can estimate the variance of this distribution. The relationship between the variance of the means of samples and that among individual observations can be shown as

$$\sigma_{\bar{X}}^2 = \sigma^2/n \qquad \text{or} \qquad s_{\bar{X}}^2 = s^2/n$$

for the *variance of the mean* and,

$$\sigma_{\bar{X}} = \sqrt{\sigma^2/n} = \sigma/\sqrt{n} \qquad \text{or} \qquad s_{\bar{X}} = \sqrt{s^2/n} = s/\sqrt{n}$$

for the *standard deviation of the mean* (often called *standard error* or *standard error of the mean*), where n denotes the number of observations comprising the sample with mean $\bar{X}_i$. The frequency distribution of the sample means has the same mean, μ, as the original population from which all the observations were drawn but it has a smaller variance (only $1/n$ as large) and a smaller standard deviation (only $1/\sqrt{n}$ as large).

As will soon be shown, just as from a single sample mean $\bar{X}_i$ one can estimate the variance of individuals, from a group of sample means one can compute the variance in the population of $\bar{X}_i$'s. It is common practice among researchers to publish the value of a sample mean, plus or minus the standard error of the mean ($\bar{X} \pm s_{\bar{x}}$). This gives the reader an idea as to how much variability one would expect to find in the means of many samples of size n drawn from the same population.

These concepts regarding dispersion are illustrated in Table 2-4, where are presented data of an experiment in which the effectiveness of a new antibiotic drug was tested on a particular strain of bacteria.

The *coefficient of variation*, CV, is another method of measuring dispersion. It is not intended as a substitute for either the variance or the standard deviation of the sample. Rather, it is intended for use on those occasions when one wants to make a comparison of the variability in samples which have substantially different means and perhaps are even measured in different units where comparisons of s^2's or s's are clearly inappropriate, To cite a ludicrous but perhaps useful example, say that one wished to compare the variability in sample weights of apples

Table 2-4 The results of a test of a new antibiotic drug on a strain of bacteria [a]

Sample number	Number resistant colonies per sample	$X_i - \bar{X}$
1	28	−5.4
2	30	−3.4
3	26	−7.4
4	42	8.6
5	30	−3.4
6	28	−5.4
7	52	18.6
8	32	−1.4
9	40	6.6
10	26	−7.4
Total	334	0

$$\bar{X} = \frac{\sum X_i}{10} = \frac{334}{10} = 33.40$$

$$s^2 = \frac{\sum(X_i - \bar{X})^2}{n-1} = \frac{\sum X_i^2 - (\sum X_i)^2/10}{10-1} = \frac{11{,}812 - (334)^2/10}{9} = 72.93$$

$$s = \sqrt{s^2} = \sqrt{72.93} = 8.54$$

$$s_{\bar{x}}^2 = s^2/n = 72.93/10 = 7.29$$

$$s_{\bar{x}} = \sqrt{s_{\bar{x}}^2} = \sqrt{7.29} = 2.70$$

$$\bar{X} \pm s_{\bar{x}} = 33.40 \pm 2.70 = 30.70 \text{ to } 36.10$$

[a] Each sample consisted of a culture plate inoculated with a fixed number of cells. Resistant colonies arose from cells resistant to the effects of the drug.

(measured in grams) and those in anteaters (measured in pounds). The coefficient of variation for each sample is calculated as

$$CV = \frac{100s}{\bar{X}}$$

where $\bar{X}$ and s are the familiar sample statistics. This value measures variability (by means of s) as a percentage per mean unit. CV is a dimensionless measure of variability; it follows, then, that variations in samples recorded in grams and pounds can be compared with complete ease. Likewise, because of their relative sizes, one naturally expects the differences in weights among anteaters to be greater than those among apples. So, if one chose to compare the s's of the two samples and found the ratio to be, say, 30 : 1 (anteaters to apples) it would have little functional meaning whereas a comparison of the dimensionless CV's might be more useful.

More realistic examples of application of this statistic might be found in comparisons of variability among numbers of red and white blood cells in samples where the means of the two cell types in samples typically vary by a factor of 400, or perhaps in the case of samples drawn from different experiments but involving the same characters.

Coding

Frequently calculations can be made considerably easier by the procedure of *transcribing* the original observations into a new scale; that is, by *coding*. Coding is most often done when the original values of the variables contain just too many digits and are, therefore, cumbersome. While just about any type of coding scale may be applied to the original data, some care must be exercised in following the rules of decoding so that the meaning of the statistic is not altered.

Additive coding consists of adding or subtracting a constant from each observation, so that

$$X_1 = 468 \qquad X_2 = 473 \qquad X_3 = 459$$

might be coded by subtracting 400 from each X_i to produce

$$X_1' = 68 \qquad X_2' = 73 \qquad X_3' = 59$$

Obviously the sample mean, $\bar{X}$, will change as a function of the constant that has been added to or subtracted from the observations but it can readily be decoded by performing the reverse operation. For the data of this example

$$\bar{X} = 466.7 \qquad \bar{X}' = 66.7$$

and

$$\bar{X} = \bar{X}' + 400$$

However, the variance and standard deviation are not affected by this type of coding since adding or subtracting a constant does not change sample dispersion, so that

$$s^2 = 50.4 = s^{2\prime}$$

and

$$s = 7.1 = s'$$

for this example.

In *multiplicative coding* the mean is again affected, but in this case, too, it can be decoded by performing the reverse operation. For example, the following observations

$$X_1 = .172 \quad X_2 = .301 \quad X_3 = .206$$

might be coded by multiplying each value by 100 so that

$$X_1' = 17.2 \quad X_2' = 30.1 \quad X_3' = 20.6$$

with

$$\bar{X} = .226 \quad \text{and} \quad \bar{X}' = 22.6$$

and

$$\bar{X} = \bar{X}'/100$$

The variance and standard deviation are affected in this type of coding, but each is altered in a different fashion. For example, for these data

$$s^2 = .00447 \quad s^{2\prime} = 44.7$$

and

$$s^2 = s^{2\prime}/100^2$$

The variance has to be divided by the multiplicative code, squared, since the terms of the equation are squared and the coding factor becomes squared, too. However, in the case of the standard deviation

$$s = .067 \quad s' = 6.7$$

and

$$s = s'/100$$

Combination coding combines both additive and multiplicative coding in the same observation so that both earlier noted sets of rules of decoding must be applied. For example, if the observations have been coded as

$$X_i' = X_i \times 100 + 4.8$$

the coded mean can be restored to its true value as

$$\overline{X} = (\overline{X}' - 4.8)/100$$

The variance is restored as

$$s^2 = s^{2\prime}/100^2$$

and the standard deviation as

$$s = s'/100$$

as the rules of additive coding can be ignored for these latter two statistics.

PROBLEMS

1. The weights of 25 immature male mice from a common laboratory strain were recorded (gm).

56.4	60.2	67.4	70.8	64.2	71.4	60.9	69.8	72.2
71.4	65.2	63.6	58.2	64.1	68.9	55.2	70.3	61.9
71.4	63.9	55.3	70.5	60.2	66.5	58.1		

From these data (a) Calculate the mean of the sample. (b) Determine the modal class. (c) Determine the median class. (d) Calculate the variance of the sample. (e) Calculate the variance of the sample mean. (f) Calculate the standard deviation of the sample. (g) Calculate the standard error of the mean. (h) Calculate the coefficient of variability. (i) Code the data by subtracting 50 from each entry. Calculate the mean of the coded data; then decode it and compare your answer to that of part (a) above.

2. Reexamine the data of Table 2-4. Consider the column *Number resistant colonies per sample* and perform the stated coding operations, after which, decode and compare to the original values.
(a) Subtract the number 20 from each value and compute the coded mean and variance. (b) Multiply each value by 0.5 and compute the coded mean, variance, and standard deviation of the samples. (c) For each value multiply by 0.5 and subtract 10. Compute the coded mean, variance, and standard deviation of the samples.

ANSWERS TO PROBLEMS

1. (a) $64.72 = \overline{X}$
   (b) 71.4 = modal class
   (c) 64.2 = median class
   (d) $31.09 = s^2$
   (e) $1.24 = s_{\bar{x}}^2$
   (f) $5.57 = s$
   (g) $1.11 = s_{\bar{x}}$
   (h) $1.72 = CV$
   (i) $1.24 = \overline{X}'$

3
PROBABILITY REVISITED

Probability distributions

In the previous chapters we were concerned with both the probability of occurrence of an event and the relative frequency of its observed occurrence in a sample. On the one hand, it was seen that the calculation of the probability of occurrence was founded on the perhaps unreal basis of a sample with an infinite number of observations, while the observed frequency of occurrence could be described by a relative frequency distribution based on a sample with a finite number of observations, like that shown in Figure 2-1. It is a simple matter to combine these two concepts into that of the *probability distribution*, which describes the relative probability of the occurrence of *an* observation based on the relative frequency distribution of an infinitely large *sample* of observations; that is, the limit of an observed frequency distribution containing an infinite number of observations.

Individual probability distributions are described by formulas. To cite a simple example, consider the probability distribution for the tosses of a die

$$P(X = X_i) = 1/6; \quad X_i = 1, 2, \ldots, 6$$

This is read as "the probability, P, that an observation, X, can take on a particular value, X_i, equals 1/6." This equation can be interpreted further to say that any one face of a die will turn up with the same probability as any other face, and that each could be expected to occur exactly 1/6 of the time, if sample size were infinite. As might be expected, there are different types of probability distributions. (Each can be visibly represented by a graph if desired, but this is by no means necessary and, in general, the use of graphs in this application is restricted to that of illustration.)

One type of probability distribution is called *discrete* because it describes the probability of observing a discrete event, like the number of dots on the face of a die. When graphed, this type of distribution is usually depicted by a bar graph (more correctly called a *histogram*), as shown in the hypothetical example of Figure 3-1, where the probability of occurrence of the event can be read off directly. All of the class intervals are of equal width although the heights vary. As the total area under the probability distribution curve equals unity (the sum of the

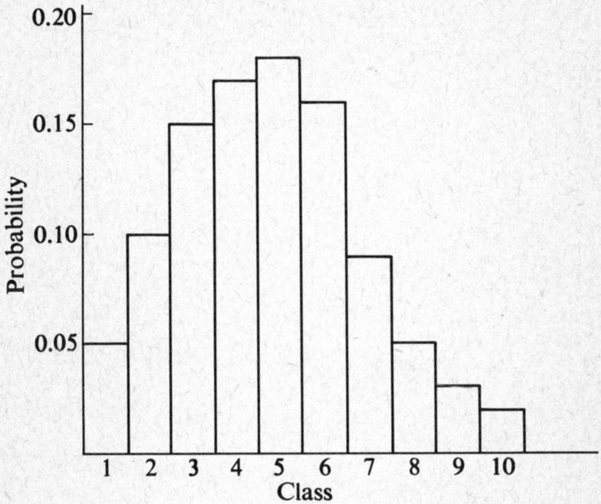

Figure 3-1 Hypothetical discrete probability distribution.

probabilities of all possible occurrences of a variable must always equal 1.00), it follows that the probability of any one occurrence must be proportional to the *area* (length × width) enclosed by a given rectangle. The area enclosed by the class 4 rectangle, for example, is proportional to the probability of observing a class 4 event. Thus, the probability of observing a value greater than 4 is proportional to the area under the curve to the right of class 4; that is

$$P(X > 4) = \sum P(X = X_i) \qquad X_i = 5, 6, 7, \ldots, 10$$
$$= .18 + .16 + \cdots + .02$$

Likewise, the probability of observing a value less than 4 includes the portion to the left; that is

$$P(X < 4) = \sum P(X = X_i) \qquad X_i = 1, 2, 3$$
$$= .05 + .10 + .15$$

The second main type of probability distribution is called *continuous* since it is applied to continuous variables. Though it was earlier noted that observations made on continuous variables can never be perfectly described due to the impossibility of making an infinitely refined measurement, the continuous probability distribution, being based on an unreal situation, can. Specifically, *discrete intervals* can be set up around any *continuous measurement*, with the width of the interval being set at any convenient value. Let us say that one is interested in setting up a probability distribution with weight as the variable of measurement where the range of observations is, say, approximately 11 gm (i.e., 4–15 gm) and the probability of observing a weight less than 5 or greater than 14 gm is very small. One may choose to set up the intervals as shown in Figure 3-2 where the total area under this curve equals 1.00 while that under the curve and within the interval between, say, 5 and 6 gm denotes the probability of observed weight falling in the interval 5.0000 ... to 5.9999 ... gm. The probability of observing a value $>5.9999\ldots$ is equivalent to the area under the curve to the right of 6, and a value $<5.0000\ldots$ the area to the left of 5.

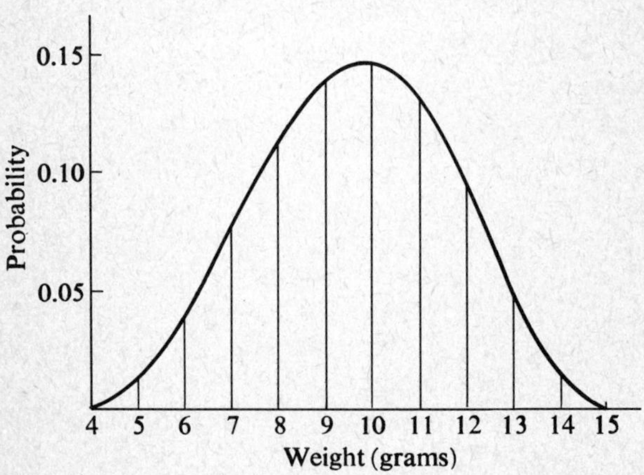

Figure 3-2 Hypothetical continuous probability distribution. Note that although the variable [grams (weight)] is continuous, the intervals are discrete.

In analogous fashion probabilities of occurrence can be found for any interval. For example, the probability of observing an event lying in the interval 4.5000 ... to 4.7999 ... can be calculated by computing the area under the curve in this interval. However, since the area under a curve describing a continuous variable can often be found only by mathematical integration, which can be tedious, *cumulative probability distributions* for many continuous variables have been computed by mathematicians. Using such a distribution, the probability of observing a value in any interval can be read off directly from the table. As an example of a cumulative probability distribution, we can apply a wheel of fortune like that seen at carnivals and depicted in Figure 3-3. For

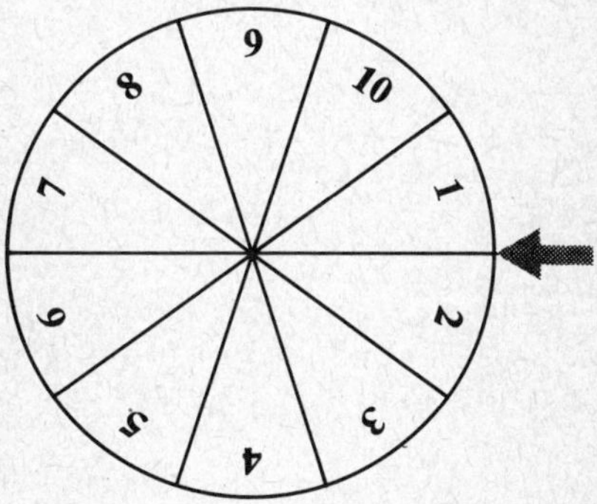

Figure 3-3 A wheel of fortune.

the purpose of simplicity, in this example the wheel will initially be divided into only 10 sectors, or intervals, although there is no reason why it could not be divided into 100 or even 100 million since the pointer can lie at any point on the wheel. Since the pointer if made increasingly thin could point to increasingly small divisions of the wheel, the divisions of the wheel represent a *continuously distributed variable*. However, regardless of the size and number of divisions, the total area—i.e., total probability—included by the wheel remains fixed at 1.00. The probability of the pointer stopping within any one of the 10 sectors can be described by the formula

$$P(X = X_i) = 1/10 \qquad X_i = 1, 2, 3, \ldots, 10$$

The range of possible values taken on by X may also be written

$$1 \leq X \leq 10$$

which states that "the value of an interval X can lie between $X_i = 1$ and $X_i = 10$, inclusive." Let us say, however, that it is not sufficient to know just the probability of the pointer stopping in any one of the 10 intervals but that it is also necessary to know the probability of its stopping within any interval. This information can be obtained from the cumulative probability distribution associated with this system, which is graphically depicted in Figure 3-4. In this figure the total area

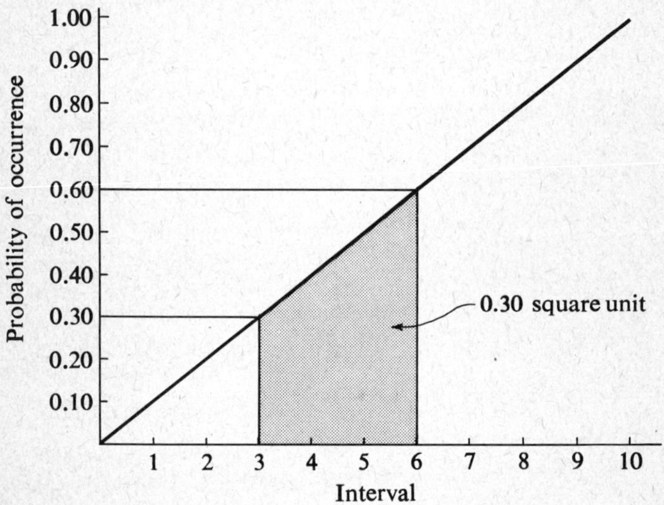

Figure 3-4 Cumulative probability distribution for a wheel of fortune.

under the curve (a sloping line in this case) is equal to 1.00 since it includes the probabilities of occurrence of all possible events. Assume that one wishes to find the probability that the pointer will stop somewhere between, say, intervals 3 and 6; that is,

$$P(3 \leq X \leq 6) = ?$$

First, one reads up from $X = 3$ to the sloping line and then over to the Y-axis to obtain the probability of the pointer stopping somewhere in the 1st, 2nd, or 3rd interval, which is found to be $P = 0.30$. This is repeated for $X = 6$, where the value $P = 0.60$ is obtained. Now, the probability of the pointer stopping in the first 3 intervals is subtracted from the probability of its stopping in the first 6 intervals, so that the whole problem can be summarized as

$$P(3 \leq X \leq 6) = P(X \leq 6) - P(X \leq 3) = 0.60 - 0.30 = 0.30$$

If one wished a finer measure of probability, it would be a simple matter to divide each interval into, say, tenths. Then, to find the probability of the pointer lying in some interval between 2.7 and 5.8, for example, one would calculate

$$P(2.7 \leq X \leq 5.8) = P(X \leq 5.8) - P(X \leq 2.7) = 0.58 - 0.27 = 0.31$$

Some selected probability distributions

Thus far our only really specific examples of probability distributions have dealt with rather simple, almost intuitively obvious situations like coin tossing games and wheels of fortune. However, it is a decidedly different matter if, for example, one desires to describe a probability distribution for a more realistic biological variable like estrogen level in month-old female mice, when perhaps all one has to work with is a frequency distribution from a sample that may not be particularly large and that may not even be a particularly good representative of the population. Fortunately, most biological variables can be classified as belonging to one of a relatively small number of mathematical distributions. Once the general type of distribution to which a variable belongs is determined, the equation describing the distribution is fixed and it is then a relatively simple matter to examine the observations within the framework of the expectations. We can now begin an examination of some of the most commonly encountered mathematical distributions.

The *binomial distribution* was first encountered in Chapter 1, though it was not labeled as such, and denotes a two-possible event—that is, a "binomial" system. The expression describing this particular distribution can be written

$$(p_1 + p_2)^n$$

where p_1 = probability of occurrence of event 1,
p_2 = probability of occurrence of event 2, and
n = number of trials or observations in a sample.

For a case such as predicting the array of sexes of children in a family—clearly a binomial system—one can let

p_1 = probability of observing a male = 0.50,
p_2 = probability of observing a female = 0.50, and
n = number of children in the family.

Assuming a family of, say, 3 children

$$(p_1 + p_2)^3 = p_1^3 + 3p_1^2 p_2 + 3p_1 p_2^2 + p_2^3$$
$$= \left(\frac{1}{2}\right)^3 + 3\left(\frac{1}{2}\right)^2\left(\frac{1}{2}\right) + 3\left(\frac{1}{2}\right)\left(\frac{1}{2}\right)^2 + \left(\frac{1}{2}\right)^3$$

In that $p_1 = p_2 = 0.50$ and $n = 3$, the result of this equation may be interpreted as the probability of observing

a. 3 males $= p_1^3 = 0.50^3 = 0.125$,
b. 2 males and 1 female, in any order $= 3p_1^2 p_2 = 3(0.50)^2(0.50) = 0.375$,
c. 1 male and 2 females, in any order $= 3p_1 p_2^2 = 3(0.50)(0.50)^2 = 0.375$,
d. 3 females $= p_2^3 = 0.50^3 = 0.125$.

The equation describing the binomial distribution can be expanded to include *any* values of p_1, p_2, and n with the provisos that $p_1 + p_2 \equiv 1.00$ and n must be a positive integer—that is, a positive whole number.

The most difficult part of calculating various expansions of the binomial is the determination of the coefficients of the various terms for a relatively large value of n. However, they can be readily found by using a simple but most effective system called *Pascal's Triangle*, shown below for $1 \leq n \leq 6$. One can easily determine how it is derived:

n	Coefficients
1	1 1
2	1 2 1
3	1 3 3 1
4	1 4 6 4 1
5	1 5 10 10 5 1
6	1 6 15 20 15 6 1

except for the 1's at the outer edges, each number is the sum of the two numbers just above, as illustrated by those numbers in the dashed triangle.

A different form of the binomial formula is used when a knowledge of the entire array of binomial events is not needed; rather just one outcome such as, for example, the probability of observing 2 males in a family of 3 children. In this formula the probability that a "favorable" event (e.g., a male child) will be observed exactly X times ($X = 2$ in this example) in families of $n = 3$ children is

$$P_X = \frac{n!}{X!(n - X)!} p^X(1 - p)^{n-X}$$

$$P_{x=2} = \frac{3!}{2!(3 - 2)!} 0.5^2(1 - 0.5)^{3-2}$$

The symbol "!" denotes a *factorial* instruction and is a shorthand way of writing the *product of a series* like

$$X! = (X)(X - 1)(X - 2) \cdots (1)$$

and where by definition $0! = 1$.

A binomial probability distribution for, say, the occurrence of X sons in families of n children or the occurrence of X Heads in n tosses is special in that it is symmetrical; that is, while the probability of any one event being favorable (p) equals 0.50, the probability of the event not being favorable ($1 - p$) also equals this value. However, in most cases the distribution is skewed to one side as in the earlier example regarding the nucleotide Adenine and its probability of occurrence as described in Table 1-1. Here there are 4 possible events (one for each nucleotide) but only one satisfies the condition of Adenine, so that the probability of the favorable event, p, equals 0.25 and the probability of not obtaining the favorable event equals $1 - 0.25 = 0.75$. Thus, to determine the probability of observing exactly one Adenine in a sample of $n = 3$ nucleotides one computes

$$P_{x=1} = \frac{3!}{1!(3 - 1)!} 0.25^1(0.75)^{3-1} = \frac{3 \cdot 2 \cdot 1}{1(2 \cdot 1)} 0.25(0.75)^2$$

$$= 0.42$$

If one wished to determine the probability of observing *at least* one Adenine in a sample of 3 nucleotides, one would also have to compute the probabilities $P_{x=2}$ and $P_{x=3}$, which are added to $P_{x=1}$ since all three alternatives satisfy the requirement.

The *normal* distribution (also called the Gaussian curve) is biologically the most important of all probability distributions since its bell-shaped, symmetrical curve† nicely describes the majority of variables—especially continuous ones—encountered in the biological sciences. In addition, it approximates the binomial distribution as well as the frequency distributions of some sample statistics like $\bar{X}$ and s^2 when the number of observations is large.

The formula (shown below) for the normal curve may initially appear complicated but, as will be seen, it really is quite easy to apply in a special modified form to be discussed shortly.

$$Y_i = \frac{1}{\sqrt{2\pi}\,\sigma}\, e^{-(X_i-\mu)^2/2\sigma^2}$$

where $\pi = 3.1416\ldots$,
$e = 2.718\ldots$, the base of natural logarithms,
$\mu =$ the mean of the population of variables,
$X_i =$ the value of an observed variable, and
$Y_i =$ the Y-coordinate of the point on the curve, $X_1 Y_1$.

Like the binomial distribution there are an infinite number of normal distribution curves possible since their shapes *depend entirely* on the values of μ and σ, which are themselves infinitely variable. However, when μ and σ are not known, their sample estimators, $\bar{X}$ and s, are used. Figure 3-5 illustrates three normal distributions whose shapes and/or positions on the abscissa are different though they vary only in regard to their values of μ and σ. It can be seen that μ determines the position of the curve on the X-axis, whereas σ determines the degree of dispersion of the curve—(i.e., its degree of spread and, consequently, its height). The total area included by each of the curves is identical.

While it is comparatively easy to find the probability of 10 successes in 20 trials using the binomial formula, one would have a hard time determining the probability of an observation being equal to, say, precisely 1.0350 ... mm in a normal distribution, since length is a continuous variable and the probability of observing this exact value is extremely small. However, an interval can be set up around this value

---

† While all normal distributions are somewhat bell-shaped and symmetrical, not all bell-shaped, symmetrical distributions are normal. For normality to obtain, the curve must be describable by means of the formula of the normal curve.

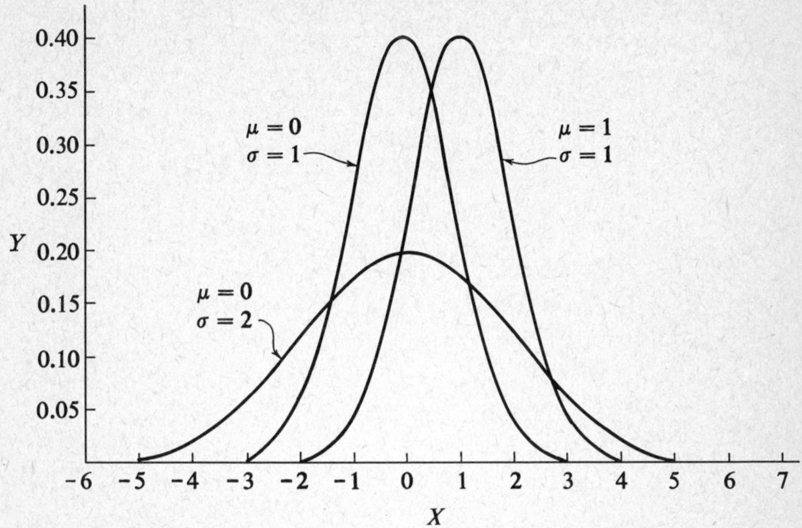

Figure 3-5 Three normal distributions. (See text for explanation.)

with the limits at, for example, 1.000 ... and 1.499 ... mm. From the area under the curve within that interval one can determine the probability of the observation lying within that interval. But this would prove to be a laborious job since first the normal curve would have to be plotted using the sample statistics $\overline{X}$ and s, and second the areas within the limits must be calculated by means of the calculus. In fact, the areas within every conceivable interval of interest would have to be individually calculated. However, there is available a standardizing measure that obviates the calculation of the normal distribution by the previously shown sample. This new measure, really a new distribution, is called the *standardized normal deviate* and is symbolized by the letter z. It is defined as

$$z = \frac{X - \mu}{\sigma}$$

and, where μ and σ are not known, it can be written using estimators, as

$$z = \frac{X - \overline{X}}{s}$$

The distribution of this variable is normal with $\mu = 0$ and $\sigma = 1$, and it has the tremendous advantage of converting *all normal distributions* regardless of value of μ and σ to a single standard z distribution with mean of zero and standard deviation equal to unity. This quantity is distributed as

$$Y_i = \frac{1}{\sqrt{2\pi}} e^{-z^2/2}$$

In the example shown in Table 3-1 the frequency distribution of weight gains, if plotted as a histogram—number of observations on midclass value—would be seen to strongly suggest distribution with

Table 3-1 The weight gains of 200 animals over a 21-day period

Class interval (gm)	Midclass value $(X_i)^a$ (gm)	Number of observations $(n_i)^a$
2.50– 7.49	5	1
7.50–12.49	10	2
12.50–17.49	15	10
17.50–22.49	20	22
22.50–27.49	25	26
27.50–32.49	30	48
32.50–37.49	35	34
37.50–42.49	40	24
42.50–47.49	45	20
47.50–52.49	50	8
52.50–57.49	55	3
57.50–62.49	60	2

[a] $\Sigma n_i = 200$
$\bar{X} = \Sigma n\bar{X}_i / \Sigma n_i = 6{,}440/200 = 32$ gm
$s = 10.1$

mean = 32 gm and standard deviation = 10.1. (The reader is asked to accept for the time without further qualification the fact that the distribution is normal.) Then by means of this sample and the z distribution, one can predict the probability of making similar observations—under similar experimental conditions, of course. One might ask, for example, the probability of observing a weight gain in excess of 50 gm—i.e.,

$P(X_i > 50)$. First, the sample observation must be converted to the z distribution by calculating

$$z = \frac{X_i - \overline{X}}{s} = \frac{50 - 32}{10.1} = \frac{18}{10.1} = 1.8$$

From the cumulative z probabilities tabulated in Table A in the Appendix one sees that $P(z > 1.8) = 0.0359$. Therefore, a little more than 3.5% of the time in similar samples one would expect an animal to gain more than 50 gm in 21 days (see Figure 3-6).

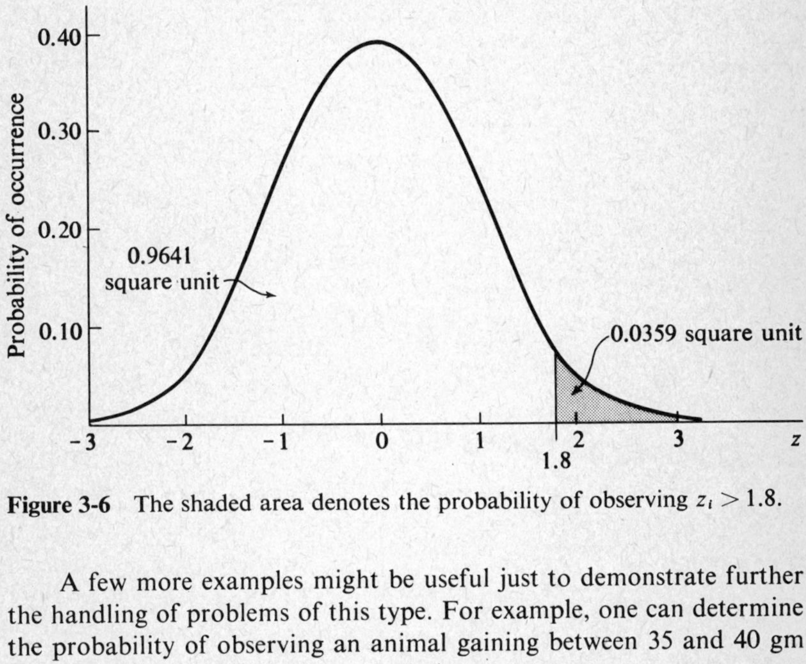

Figure 3-6 The shaded area denotes the probability of observing $z_i > 1.8$.

A few more examples might be useful just to demonstrate further the handling of problems of this type. For example, one can determine the probability of observing an animal gaining between 35 and 40 gm [i.e., $P(35 < X_i < 40)$] by first calculating

$$\begin{aligned} P(X_i > 35) &= P[z > (35 - 32)/10.1] \\ &= P(z > 0.3) = 0.3821 \end{aligned}$$

and then calculating

$$\begin{aligned} P(X_i > 40) &= P[z > (40 - 32)/10.1] \\ &= P(z > 0.8) = 0.2119 \end{aligned}$$

Thus

$$P(35 > X_i > 40) = P(0.3 < z < 0.8) = P(z > 0.3) - P(z > 0.8)$$
$$= 0.3821 - 0.2119 = 0.1702$$

Therefore, a little more than 17% of the time in similar samples one would expect an animal to gain between 35 and 40 gm in 21 days (see Figure 3-7).

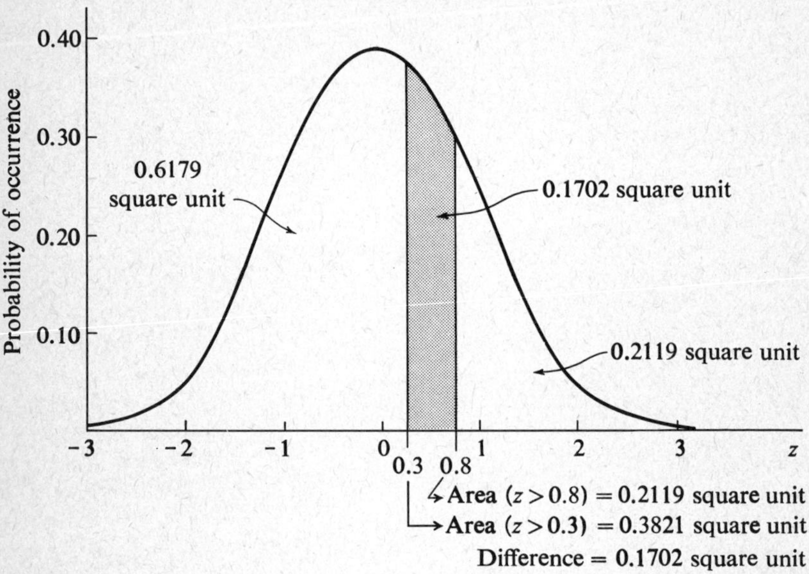

Figure 3-7 The shaded area denotes the probability of observing $0.3 < z_i <$.

Since the z distribution is symmetrical, in order to save space only the positive values of this variable are found tabulated in Table A; thus, they represent only one-half of the total area under the probability curve. However, assume that one wishes to find the probability of recording a measurement, z_i, lying in an interval that includes the population mean, μ, and therefore, lies in an interval bounded by a positive and a negative z. Consider the interval $-0.5 < z_i < 0.6$. Since the curve is symmetrical (50% of the area under the curve lies from $0 \leq z_i \leq +\infty$ and 50% from $-\infty \leq z_i \leq 0$) the probability of observing a negative z_i is the same as that for a positive one; specifically

$$P(z_i < -0.5) = P(z_i > 0.5)$$

As can be noted from Figure 3-8,

$$P(z_i < -0.5) = P(z_i > 0.5) = 0.3085$$

and

$$P(z_i > 0.6) = 0.2743$$

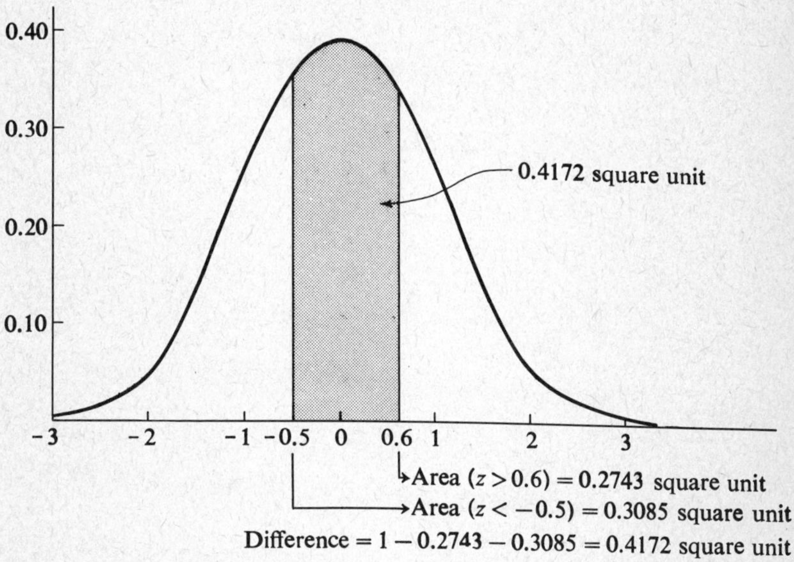

Figure 3-8 The shaded area denotes the probability of observing $-5 < z_i < 0.6$.

Then,

$$1 - 0.3085 - 0.2743 = 0.4172$$

is the probability of the observation lying within that interval.

You will note that the z table is tabulated only to $\pm z = 3.9$, beyond which points the probability of observing some z_i equals zero. However, this does not mean that one would never find a value lying outside these limits! It simply denotes that the probability of observing a value of z_i numerically greater than 3.9 is quite small.

In scientific publications one often sees the probability of occurrence of an observation expressed in *standard deviation*, or *sigma* (σ), *units*. For example, it may be stated that a "measurement of a certain magnitude has a probability of occurrence that lies within $\pm 2\sigma$ of the mean." The "σ" referred to in this context *is not* simply $\sqrt{\sigma^2}$, the square root of the variance of the population. Instead, σ and z units are synonymous, so that a σ unit refers to one standard deviation (or z) unit in a normally distributed population with mean equal to zero and standard deviation equal to unity; that is, a z distribution.

By an examination of the z distribution probabilities in Table A one can note that, for example, 95% of all observations in such a population are expected to lie within $\pm 1.96z$ of the mean, zero. Taken another way

$$P(-1.96 < z_i < 1.96) = 0.95 = \pm 1.96\sigma$$

Likewise,

$$P(-1 < z_i < 1) = 0.6826 = \pm 1\sigma$$

That is, about two-thirds of all observations can be expected to occur within $\pm 1z$ unit of the mean, and so on.

Confidence intervals

Throughout most of this text we have been concerned with the concept of statistical inference: the process by which one can infer the structure of a population from evidence obtained by sampling. Since most quantitative biological observations are more or less normally distributed, most statistical tests of inference are based on the assumption that the sample observations are taken from a normally distributed population. Although absolute conformity to normality in the population sample at one time was considered essential, it has been shown that this condition is *by no means essential* and all but the most highly skewed distributions can be handled with facility.

In Chapter 2 we saw that one could infer, or estimate, the value of μ from $\overline{X}$, σ^2 from s^2, and so on. But, often it is not adequate simply to state that $\overline{X}$, for example, equals 32.7 cm when one does not have the slightest inkling as to how close to μ it really is. If the original population is at least approximately normally distributed, one can set up an interval around $\overline{X}$ in which one can place a certain degree of confidence that

μ lies. In setting up a *confidence interval* around $\overline{X}$, the mean of a sample drawn from what is believed to be an at least approximately normally distributed population, one can convert the sample distribution to a *Student's t distribution*, which is defined

$$t_{n-1} = \frac{\overline{X} - \mu}{s_{\overline{X}}}$$

where n = sample size and
$n - 1$ = degrees of freedom, *df*.

The cumulative probabilities associated with this distribution are shown in Table B in the Appendix. The shape of the t distribution is symmetrical although there is a specific distribution for every value of degrees of freedom and, as these values approach infinity (that is, $df > 120$), the t distribution itself approaches normality. The value of t may be positive *or* negative depending on whether $\overline{X}$ or μ is the larger quantity. The probabilities associated with the various values of t are handled like any of the previously noted probability distributions. For example, for $df = 10$ Table B shows that 95% of the values are expected to lie within the interval $t = \pm 2.228$; that is

$$P(-2.228 < t_i < 2.228) = 0.95$$

for $df = 10$. This statement can be rearranged and rewritten as

$$P(\overline{X} - t_{n-1}\, s_{\overline{X}} < \mu < \overline{X} + t_{n-1} s_{\overline{X}}) = 0.95$$

where t_{n-1} is the tabular value of t for $n - 1$ degrees of freedom at the 95% confidence level. This latter equation may be interpreted as stating that the probability of μ being found in the interval $\overline{X} \pm t_{n-1} s_{\overline{X}}$ is 95%; conversely, the probability of μ lying outside the interval equals 5%.

As an example of this calculation we can use the data of Table 3-1 where

$$n = 200$$

$$n - 1 = 199$$

$$\overline{X} = 32 \text{ gm}$$

$$s = 10.1$$

$$s_{\overline{X}} = 0.7$$

The 0.95 confidence interval will be of the form

$$32 \pm t_{199}(0.7) \quad \text{or}$$
$$32 \pm 1.96(0.7)$$

which becomes

$$32 - (1.96)(0.7) < \mu < 32 + (1.96)(0.7)$$

so that one can state with 95% confidence (i.e., one is confident that 95% of the time in similar samples) μ will be found to lie in the interval 30.6 to 33.4 gm.

One can also set up a confidence interval about a sample variance, assuming that s^2 comes from an at least approximately normal distribution. The test used is based on a distribution called the *chi square* (χ^2), which is a probability distribution which itself approaches normality with $\mu = 0$ and $\sigma = 1$ when the degrees of freedom are very large. This distribution will be discussed in much greater detail in Chapter 10. However, for the time being we will simply note that one can set up an interval about s^2 in which one may have a certain degree of confidence that σ^2 lies.

By definition

$$\chi^2 = \frac{(n-1)s^2}{\sigma^2}$$

where $n - 1$ denotes degrees of freedom. Clearly, a χ^2 value can only be positive since all terms are positive. One can write

$$P(\chi^2_{.025} < \chi^2_i < \chi^2_{.975}) = 0.95$$

where $\chi^2_{.025}$ denotes the value for $n - 1$ degrees of freedom below which 0.025 of the distribution lies $[P(\chi^2_i < \chi^2_{.025}) = 0.025)]$ and $\chi^2_{.975}$ that value below which lies 0.975 of the distribution $[P(\chi^2_i < \chi^2_{.975}) = 0.975]$. Thus, using the relationship

$$P(\chi^2_{.025} < \chi^2_i < \chi^2_{.975}) = 0.95$$

and the definition formula, one can derive the confidence interval around σ^2 such that

$$P\left[\frac{(n-1)s^2}{\chi^2_{.025}} < \sigma^2 < \frac{(n-1)s^2}{\chi^2_{.975}}\right] = 0.95$$

For the data of Table 3-1 where $n - 1 = 199$ and $s^2 = 101.4$ the 0.95 confidence interval about σ^2 is found as

$$P\left[\frac{199(101.4)}{83.3} < \sigma^2 < \frac{199(101.4)}{40.5}\right] = 0.95$$

and

$$P(242.2 < \sigma^2 < 498.2) = 0.95$$

where the χ^2 values for the appropriate degrees of freedom are obtained from Table C.

By consulting this table and substituting the proper values in the equation, an interval can be set up for any degree of confidence; a 90% interval, for example, requires the substitution of $\chi^2_{.05}$ and $\chi^2_{.95}$. Once the confidence interval for the variance is computed it can be converted to one for *standard deviation*, σ, simply by taking the *square roots* of the confidence values about σ^2.

Sampling error

It has often been noted that randomness of the observations drawn from any population is required. While it is obviously hoped that each $\overline{X}$ is a reasonably accurate estimate of μ, it would be only under the most fortuitous circumstances that $\overline{X}$ would, indeed, equal this parameter. This discrepancy between any $\overline{X}$ (or X_i, for that matter) and μ is due to what is called *sampling error*. This expression does not imply that a mistake has been made but refers to the fact that of all possible values that can be drawn from a population, the very vast majority will not equal μ. We know that one overcomes sampling error by increasing sample size, but how is this accomplished? The *linear additive model* attempts to explain this system by assuming that each X_i would equal μ precisely were it not for the random error term, ε (the Greek letter "epsilon"). That is

$$X_i = \mu + \varepsilon_i$$

where the ε_i's are assumed uniquely drawn for each X_i from a normally distributed population of ε's with $\mu_\varepsilon = 0$. Since this distribution is symmetrical about $\mu = 0$ there is as great a likelihood of drawing a $+ \varepsilon_i$ as a $- \varepsilon_i$, so that if enough observations (either X_i's or ε_i's depending on your terminology) are drawn, the positive and negative error terms will cancel each other out and

$$\overline{X} = \frac{\sum_{i}^{n=\infty} (\mu + \varepsilon_i)}{n} = \mu + \frac{\sum_{i}^{n=\infty} \varepsilon_i}{n} = \mu + 0$$

Consequently, as *n* for any sample increases, one would expect the error term to decrease due to the positive and negative ε_i's canceling each other out, though most likely not to zero.

PROBLEMS

1. Define and distinguish between (a) discrete variable and continuous variable; (b) discrete, continuous, and cumulative probability distribution.
2. Set up a graph describing a cumulative probability distribution for the tossing of a single die. (*Hint:* See Figure 3-4.)
3. The probability of occurrence of twin pregnancies among all pregnancies for white females in the United States is 1 percent. While twinning does show strong familial tendencies, for purpose of illustration only we will consider it to be a random event. Among randomly selected women each of whom has had 6 full-term pregnancies, predict the frequency of women with 0, 1, 2, and 3 sets of twins.
4. In the United States the frequency of twinning among nonwhite (mostly Negro) females is about 1.4 percent. Using the same assumptions as in the previous question predict the frequency of women with 0, 1, and 2 sets of twins and compare your answer to that of Problem 3.
5. Let a "favorable" event be the occurrence of a 1 or a 2 on the face of a die. Since the occurrence of any one face on a fair die is a random event with probability of occurrence of 1/6 (approximately 0.17), compute the predicted frequency of occurrence of favorable and "nonfavorable" events (faces 3 through 6) in a series of 3 tosses.
6. A normal population has a mean of 25 and variance of 16. (a) What is the probability that in a sample randomly drawn from this population will be found a value, $X_i > 28$? (b) What is the probability that such a sample will contain an observed value $X_i > 43$? (c) What is the probability of an observed value lying in the interval, $18 < X_i < 27$?
7. Describe the 0.95 confidence interval about μ as inferred from a sample of 20 observations with a mean of 20 and variance of 25. Interpret your result.
8. Describe the 0.95 confidence interval about σ^2 using the same sample as described in Problem 7. Explain the significance of the interval.

ANSWERS TO PROBLEMS

3. $P_{x=0} = 0.941$
   $P_{x=1} = 0.057$
   $P_{x=2} = 0.001$
   $P_{x=3} = 0.000+$
4. $P_{x=0} = 0.919$
   $P_{x=1} = 0.078$
   $P_{x=2} = 0.003$
5. $$(p+q)^n = (0.34 + 0.66)^3 = \underbrace{0.039}_{3F} + \underbrace{0.229}_{2F:1NF} + \underbrace{0.445}_{1F:2NF} + \underbrace{0.288}_{3NF}$$
6. (a) $P(z > 2) = 0.0228$
   (b) $P(z > 0.75) = 0.2266$
   (c) $P(z < -1.75) = P(z > 1.75) = 0.0401$
   $P(z > 0.5) = 0.3085$
   $P(-1.75 < z < 0.5) = 1 - 0.0401 - 0.3085 = 0.6514$
7. $P(17.68 < \mu < 22.32) = 0.95$
8. $P(14.44 < \sigma^2 < 53.31) = 0.95$

4

COMPARISONS OF TWO SAMPLES

Thus far, the discussions have been limited to the analyses applied to individual samples and how they may be used to determine the general structure of the population. But now a different type of problem will be considered: namely, the comparison of two samples and the determination of whether they are both representative of the same population or, taken another way, populations with some parameters in common. In reality even if two samples are drawn from the same population, one expects them to differ because of sampling error. Then, what is really of concern in the comparison is the magnitude of the difference between the samples; from this will be judged whether or not they represent the same population. Here one must not be too rigid in the interpretation of "same population." Often it is not of concern whether the samples are drawn from the same group of individuals inhabiting, say, the same area at the same time; in fact, the comparisons of interest are often those between samples that differ spatially and/or temporally. Then, what is meant by "same population" is really better described as "populations of values with the same distribution" as measured in terms of, say, mean or variance. A sample of birthweights taken from among children born in New York City obviously cannot come from the same real population as a sample from children born in Los Angeles. But one can ask if the two groups are drawn from populations of birthweights that have a parameter in common.

It is obvious that if one looks long and hard enough at any groups of observations, some differences between them are likely to be found. In making a statistical comparison of two samples by scrutinizing the differences between them one looks for *significant differences*, and so this term must be given a specific interpretation. Significance implies that, for our particular purposes, the differences between the samples are not trivial—that is, they are real—and are, therefore, indicative of representing two different populations.

Comparisons of means

In order to begin a comparison of two samples one must first decide what hypothesis will be tested because without a hypothesis there can be no direction to the analysis. The usual one, the *null* ("zero") *hypothesis*, H_0, states that there is no real difference between the samples as regards the characteristics being compared. This hypothesis may not be correct but it is a convenient starting point. One null hypothesis might be: the two samples being compared came from populations with the same mean. This statement can be abbreviated $H_0\colon \mu_1 = \mu_2$. From this beginning the probability of various events can be deduced by means of various statistical tests to be described a little later. For the moment assuming only that the two sample means are not equal ($\bar{X}_1 \neq \bar{X}_2$), a decision will eventually be made as to whether these two $\bar{X}$'s will be judged (a) significantly different, as samples drawn from populations with different μ's, or (b) not significantly different and the difference between the two $\bar{X}$'s is no greater than that which might be found between two samples drawn randomly from populations with the same mean. If a difference of the size observed has only a very small probability of occurrence by chance, then most likely H_0 will be rejected and the *alternative hypothesis*, $H_1\colon \mu_1 \neq \mu_2$, will be accepted.

The conventional though by no means inflexible rejection point for H_0 is at the 5% level of significance. This concept can perhaps be best described by means of some examples. One can ask, for example, how often in comparisons of samples *drawn from the same population* would one expect chance alone to produce a difference between their means as large as the one observed? If a difference of this magnitude could be expected to occur in 9 out of 10 sample comparisons ($P = 0.90$) drawn from the same population, then undoubtedly this particular observed difference would be dismissed as a chance deviation attributable most likely to sampling error. Consequently, the null hypothesis of the samples being drawn from populations with the same means would be accepted. But, if a difference between sample means is so great as to be expected to occur by chance alone only once in, say, 20 comparisons of samples drawn from the same population, one might become skeptical of the notion that the two sample means in question are really representative of the same population of μ's. Even though skepticism does not constitute a proof, the doubt of the role of chance in this comparison is so great that most experimenters at this point would observe the one-out-of-twenty (5%) convention and reject the null hypothesis.

This concept may be stated as a general rule: *If the probability of occurrence of an event* (the event, here, referring to the magnitude of the difference between two sample means) *is* ≤ 0.05 *the null hypothesis is rejected*; *if* >0.05 *it is accepted.* One must never lose sight of the fact that a hypothesis can never really be proven or disproven regardless of the rejection point, as it is always possible to reject H_0 when it really is correct or, conversely, to accept H_1 when it is really incorrect. Rather, the best that one can do is to state the probability of occurrence of the observation and draw a conclusion based on that probability and the biological interpretation of the data.

One of the most widely used tests of comparison of two samples involves a comparison of their means. The quantity which best converts the difference between two sample means to a probability distribution is the now familiar Student's t distribution, as explained in the next section.

Unpaired observations

Two samples are drawn from populations which may or may not have the same μ's. For example, population A may represent one type of treatment applied to the experimental units and population B, another. Unpaired observations in such a context must meet the requirement that the ith observation in sample A, for example, is no more related to the ith observation in sample B than it is to the jth, kth, or nth in sample B. Consequently, each sample is drawn completely independently of the other. For unpaired observations

$$t = \frac{(\overline{X}_1 - \mu) - (\overline{X}_2 - \mu)}{s_{\bar{d}}}$$

$$= \frac{\overline{X}_1 - \overline{X}_2}{s_{\bar{d}}}$$

when $H_0 : \mu_1 = \mu_2$ is true. The t distribution for unpaired observations has $n_1 + n_2 - 2$ degrees of freedom. It is assumed that the populations from which the samples have been drawn are approximately normally distributed and have a common but not necessarily known variance, σ^2. The value of t can be seen to measure the magnitude of the difference between the two sample means, $\bar{X}_1 - \bar{X}_2$, divided by the standard deviation of the difference, $s_{\bar{d}}$. In general, the larger the difference between the means the greater will be the t value; thus, the likelihood of similarity of $\bar{X}$'s is inversely proportional to the value of t. The statistic $s_{\bar{d}}$ is only an estimator of $\sigma_{\bar{X}_1 - \bar{X}_2}$, the standard deviation of the true difference between the means; therefore, t is subject to sampling variation. But this estimate can be refined if the variance assumed common to both populations, σ^2, is estimated by means of a quantity called *pooled variance*, defined as

$$s_p^2 = \frac{s_1^2(n_1 - 1) + s_2^2(n_2 - 1)}{(n_1 - 1) + (n_2 - 1)}$$

where s_1^2 and s_2^2 denote the variances of the samples containing n_1 and n_2 observations, respectively.

One might wonder why the calculation of t requires the difference, $\bar{X}_1 - \bar{X}_2$, to be divided by a measure of the variability of the difference. Very simply, the variability in the samples is of major importance in determining significance. Consider Figure 4-1, for example. In part (a) we see the frequency distributions of two large samples, their means, $\bar{X}_1$ and $\bar{X}_2$, and the difference between them, d. Note that the two samples overlap to such an extent that we probably would not be too

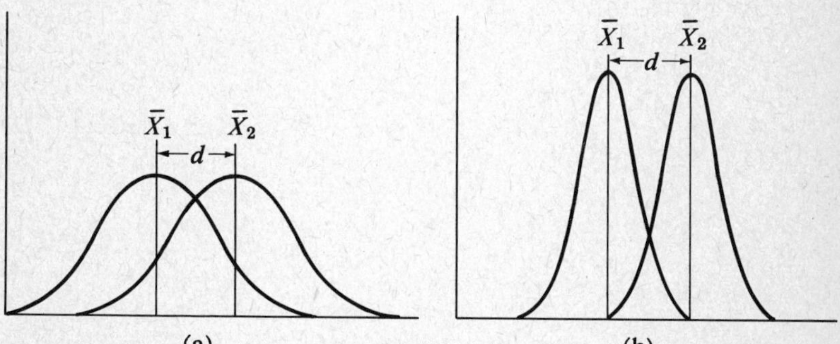

Figure 4-1 (a) Two hypothetical samples with dissimilar means but a common, relatively large, variance. (b) The means are the same as in (a), but the common variance is somewhat reduced.

surprised if told that both samples, and thus their $\bar{X}$'s, had come from the same population. On the other hand, consider part (b), where d is of the same magnitude as in (a). Yet, in this case, it seems intuitively† less likely that the two samples have been drawn from the same population, because of their relative lack of overlap.

The pooled variance is a weighted mean of the sample variances when $n_1 \neq n_2$. In this instance it is a better estimator of σ^2 than the arithmetic mean variance, $(s_1^2 + s_2^2)/2$, which weighs the two sample variances equally, regardless of sample size. However, the pooled and arithmetic mean variance are identical when $n_1 = n_2$. When $n_1 \neq n_2$

$$s_{\bar{d}} = \sqrt{s_p^2\left(\frac{n_1 + n_2}{n_1 n_2}\right)} = \sqrt{s_p^2\left(\frac{1}{n_1} + \frac{1}{n_2}\right)}$$

and when $n_1 = n_2 = n$

$$s_{\bar{d}} = \sqrt{2s_p^2/n}$$

The data of Table 4-1 present an application of the t test on two samples of equal size. Here, the mean weights of mycelia in two different fungal strains are compared. The observations in the two samples clearly are unpaired; that is, there is no apparent relationship between X_{1i} and X_{2i}. The hypothesis to be tested is $H_0 : \mu_1 = \mu_2$ as estimated by $\bar{X}_1$ and $\bar{X}_2$. According to the t probabilities shown in Table B, between 0.75 and 0.80 of the area under the curve lies *outside* of the limits of the value ± 0.28, at 4 degrees of freedom. In other words, one would expect to observe a value of t this large or larger between 75% and 80% of the time if an infinitely large number of similar samples drawn from the same population were compared. As a t value computed on the basis of a deviation, $\bar{X}_1 - \bar{X}_2$, of this size is relatively commonplace—it certainly occurs much more often than 5% of the time—one should accept the null hypothesis that the two samples have come from populations with the same mean, μ.

No consideration was given to the sign of t; the conditions of the particular test of a hypothesis applied here are such that the results are the same if $t = \pm 0.28$. The symmetry, or "two-sidedness," of this test is entirely due to the hypotheses tested: H_0 : no difference between the

† Intuition has a limited role in biostatistics and is introduced here solely for the sake of illustration.

Table 4-1 Dry weight (mg) of fungal mycelia in samples obtained in two strains of the same species

	Strain Number 1	Strain Number 2	
	246.3	246.2	
	239.2	247.1	
	257.3	244.9	
$\sum$	742.8	738.2	$n_1 = n_2 = n = 3$
$\overline{X}$	247.6	246.1	
s^2	83.17	2.22	

$$s_p^2 = \frac{83.17(2) + 2.22(2)}{2(2)} = 42.69$$

$$s_d = \sqrt{2(42.69)/3} = \sqrt{28.46} = 5.34$$

$$t = \frac{1.50}{5.34} = 0.28$$

$$df = 2 + 2 = 4$$

means of the populations, and H_1: the two population means are different. So, even if H_0 had been rejected and $H_1: \mu_1 \neq \mu_2$ accepted, it still would have made no difference if $\overline{X}_1 > \overline{X}_2$ or $\overline{X}_1 < \overline{X}_2$. However, one might have been interested in testing a *different* alternative hypothesis; that is, whether the weight of mycelia in Strain Number 1 is *significantly greater* than that in Strain Number 2. Here, the question is not simply whether the means are different but whether they differ such that $\overline{X}_1$ is significantly larger than $\overline{X}_2$ (i.e., $\mu_1 > \mu_2$). Posing this new question, no matter how large the difference between the sample means, if $\overline{X}_2$ had been observed to be larger than $\overline{X}_1$ the difference could never be judged significant with respect to the hypothesis now being tested! For any test of such a "one-sided" hypothesis, the t value is calculated in the same way as before but the interpretation of the probability of occurrence will be different than before. In the t table, if the probability that a value of t will fall outside the interval ± 0.28 lies between 0.75 and 0.80 it means that one half of the area (0.375 to 0.40) under the curve in this interval will fall above $+0.28$ and one half below -0.28. In a one-sided test only one of the areas is of interest

—either to the right of +0.28 or to the left of −0.28 but not both. Therefore, for any t value the probability of occurrence for a one-sided test is precisely one half that for a two-sided test. In the data of Table 4-1, had the alternative hypothesis been $\mu_1 > \mu_2$, the decision would still be that $\mu_1 = \mu_2$ (the hypothesis $\mu_1 > \mu_2$ would have been rejected) but now based on $0.37 < P < 0.40$.

Paired observations

Oftentimes experiments are designed so that the observations in the two samples to be compared have some known relationship to one another; and each observation in either sample can be compared with its counterpart in the other (i.e., X_{1i} versus X_{2i}). One example of this type of paired comparison occurs when two observations are made on each individual, but each observation is representative of a different population of values, as in the case when the same individual is treated with each of two different drugs. As might be expected there will be less extraneous variability in the results of the experiment than if each drug were tested on a different individual. Table 4-2 presents the results of an experiment where each of two antihistaminic drugs (coded A and B) was tested in the same seven individuals. While the t test used here is similar to the one used for comparing sample means of unpaired observations, the measure of the variability in the samples—as applied to the calculation of t—is different. In the paired comparison test, variability is computed from the differences between the paired observations, d, rather than from the individual observations of the sample. This is done because there is a component in each pair of observations that is unique to them; it is attributable to the pair of responses being measured in the same experimental unit—that is, individual. Here

$$t = \frac{\bar{d}}{s_{\bar{d}}}$$

where

$$\bar{d} = \frac{\sum (X_{1i} - X_{2i})}{n} = \frac{\sum d_i}{n}$$

and

$$s_{\bar{d}} = \sqrt{\frac{\sum d_i^2 - (\sum d_i)^2/n}{n(n-1)}}$$

Table 4-2 Effects of 2 antihistaminic drugs tested on the same individuals

	Subject no.	Hours of relief with antihistamine A (X_1)	Hours of relief with antihistamine B (X_2)	$d = X_1 - X_2$
	1	4.9	4.2	0.7
	2	1.3	2.1	−0.8
	3	10.9	7.4	3.5
	4	4.7	3.6	1.1
	5	6.5	7.2	−0.7
	6	4.6	1.6	3.0
	7	9.5	7.9	1.6
Sum		42.4	34.0	8.4
Mean		6.1	4.8	1.2

$$s_d = \sqrt{\frac{[(0.7)^2 + (-0.8)^2 + \cdots + (1.6)^2] - (8.4)^2/7}{7\,(6)}} = 0.62$$

$$t = \frac{1.2}{0.62} = 1.9 \quad \text{with}$$

$$df = 6$$

with degrees of freedom $= n - 1$. In this instance $H_0 : \mu_1 = \mu_2$ is accepted since the probability of observing an equivalent value of t computed from samples drawn from the same population lies between 10% and 15%. For this hypothesis a two-sided test is necessary. However, if one had instead been interested only in whether drug A was significantly better than drug B in producing a longer lasting relief—clearly a one-sided test—the associated t probability would be found to lie between 5% and 7.5%. However, since this probability is also greater than 0.05, this hypothesis would be rejected.

Comparisons of variances

To this point in the discussion of sample comparisons, the variance and standard deviation were used only in computing the t test relative to a test of the population mean. However, in sample comparisons it would be incorrect to relegate the variance of a population or sample to this relatively minor role. Often a comparison of population variances conveys information about their makeup that the means cannot. The test used, in principle, is much like a test of means in that, for example, the null hypothesis for a two-sided test is that the variances are not different ($H_0 : \sigma_1^2 = \sigma_2^2$), and for a one-sided test one variance is not significantly larger than the other. However, instead of testing the difference between the variances, their *ratio* is used. Consequently, $H_0 : \sigma_1^2 = \sigma_2^2$ is confirmed by the ratio $\sigma_1^2/\sigma_2^2 = 1$. We saw earlier that as the difference between two sample means increases, the probability that they have come from populations with unequal means generally increases; likewise, as the ratio of the variances increases over unity, the probability that they represent populations with unequal variances increases. The procedure used for calculating the probability distribution of this ratio of variances utilizes the *F distribution*, where

$$F = \frac{\sigma_1^2}{\sigma_2^2}$$

for populations and

$$F = \frac{s_1^2}{s_2^2}$$

for samples.

Since variances are always positive quantities, it follows that the value of F can vary from 0 to $+\infty$. However, in the cumulative F probability distribution shown in Table D in the Appendix, you will note that there is no value less than unity. This is done simply to save space in the table and is accomplished by defining this quantity as applied to the null hypothesis $H_0 : \sigma_1^2 = \sigma_2^2$ as

$$F = \frac{\text{larger } s^2}{\text{smaller } s^2}$$

where the numerator is always the larger of the two variances.

The data of Table 4-1 can be utilized for an illustration of this test. If the hypothesis to be tested is the null hypothesis of equality of the σ^2's, then

$$F = \frac{s^2 \text{ Strain Number 1}}{s^2 \text{ Strain Number 2}} = \frac{83.17}{2.22} = 37.46$$

with $n - 1 = 2$ degrees of freedom in the numerator and, $n - 1 = 2$ degrees of freedom in the denominator. The F probability associated with this particular test—which clearly is two sided because it makes no difference in the hypothesis which variance is the larger—lies between 0.05 and 0.07. Since the probability of observing this value of F exceeds 0.05, the sample variances are judged as being representative of populations with a common variance and the null hypothesis is accepted. However, if the alternative hypothesis to be tested had been one sided: specifically, that $s_1^2 > s_2^2$, the probability associated with the F value, 37.46, would be found to be <0.05 (it lies in the interval between 0.025 and 0.035) and the variances would be judged significantly different; the null hypothesis would be rejected.

Sample size

As has been repeatedly stressed, large samples provide the experimenter with a better idea of the makeup of a population than do small ones; at the extreme, if a sample is infinitely large the population can be described perfectly. In most instances the size of the sample is under the control of the experimenter but it is sometimes impractical from the standpoint of time and economics to make more observations than are absolutely essential to provide a given degree of sensitivity to the test of the hypothesis.

Minimum sample sizes that permit one to detect a given difference between populations of continuous variables can be determined by means of a modification of Student's t test provided that one has at least an approximate estimate of the maximum variability to be encountered in the sample. Probably the most common question regarding sample size is: How large must a sample be in order to be c percent confident that a real difference, d, between the means of the populations sampled will be judged significant in a t test at the s percent level of significance? The test used requires only that one have some general idea as to the maximum size of the variability of the measurements in question. If a variance estimate cannot be assumed from observations drawn on at least similar material, then a small "test sample" must be collected and used to provide the overall estimate of variance. The minimum size of the sample is calculated as

$$n = \frac{2s_e^2(t_c + t_s)^2}{d^2}$$

where n = the minimum number of observations sufficient to detect

$d = \mu_1 - \mu_2 =$ the smallest *true* difference that it is desired to detect between populations with unequal means,

$s_e^2 =$ the estimate of the variance for the observations in question,

$t_c =$ the value of Student's t for $(1 - c)$th level of confidence, using a one-sided test with $df = \infty$. (For 0.99 confidence one uses tabulated t corresponding to $1 - 0.99 = 0.01$; $t_{.01} = 2.326$ for $df = \infty$.) Likewise,

$t_s =$ the one- or two-sided tabulated Student's t value (depending on hypothesis) at the sth level of significance and $df = \infty$.

The data of Table 4-1 can provide two examples of application of this estimate. Suppose that one wished to be 95% confident of detecting a real difference as small as 3 mg between two populations of fungal mycelium weights using the 5% level of significance. For Strain Number 1

$$n = \frac{2[83.17\,(1.64 + 1.96)^2]}{(3)^2} = 239.5 \text{ observations}$$

and for Strain Number 2

$$n = \frac{2[2.22(1.64 + 1.96)^2]}{(3)^2} = 6.4 \text{ observations}$$

From these calculations one can note that the future samples expected to fulfill these requirements should contain no less than 240 observations for the one strain and 7 for the other. The tremendous disparity in estimated n's is due to the nearly 40-fold difference in the s^2's. As can also be seen from an inspection of this formula, as one increases the level of confidence, lowers the level of significance, or decreases the size of the desired detectable difference between μ's, n will become larger.

PROBLEMS

1. An environmentalist studied the number of particulate respiratory contaminants carried airborne in a large metropolitan area by two different methods. His intention was to determine the method that would be most sensitive in assessing the true concentration of these particles. Each value represents the number of particles collected in an hour in a given volume of air, by each of the two methods. All samples were collected from the same locale.

Hour	*Method 1*	*Method 2*
1	91	52
2	63	62
3	265	153
4	144	117
5	83	87
6	149	109
7	73	78
8	175	138
9	248	160
10	61	55

(a) Test the hypothesis $\mu_1 = \mu_2$ at the 0.05 level of significance. (b) Assume that the original hypothesis had been $\mu_1 > \mu_2$; test this hypothesis at the 0.05 level. (c) How would it be possible to test the hypothesis $\mu_2 > \mu_1$?

2. Each of twenty gerbils (a small, Mongolian rodent) of the same age and sex was randomly assigned to one of two treatment groups. Group A received a standard commercially prepared diet while Group B received one said to promote a more rapid weight gain. After ten days each animal was weighed and the increase in weight over the starting value recorded (gm).

Group A	*Group B*
9.6	11.2
10.2	9.6
7.4	6.4
10.2	5.6
8.8	7.6
8.4	11.8
12.8	7.8
8.4	10.2
5.0	6.2
5.8	10.8

Test the hypothesis that Group B gained more weight than Group A, at the 0.05 level of significance.

3. The difference between the two diet groups of Problem 2 was seen to be nonsignificant. However, one might encounter the argument that the sample sizes were not adequately large to detect a real difference. What are the minimum sample sizes required in order to be 0.95 confident that a real difference between the means of the two diet populations—if one does exist—will be judged significant at the 0.05 level of significance? Let the smallest true difference that it is desired to detect between the two populations, d, equal 1 gm.

4. Using the data of Problem 2 perform the F test appropriate to testing the hypothesis: $\sigma_B^2 > \sigma_A^2$.

ANSWERS TO PROBLEMS

1. (a) Paired comparison
   $\bar{X}_1 = 135.2 \qquad \bar{X}_2 = 101.1 \qquad \bar{d} = 34.1$
   $s_{\bar{d}} = 8.5 \qquad t = 4.03^{**}\ (P < 0.01) \qquad df = 9$
   (b) $P < 0.005$
   (c) Useless, since $\bar{X}_1 > \bar{X}_2$
2. Unpaired comparison
   $\bar{X}_A = 8.66 \qquad \bar{X}_B = 8.72 \qquad s_A^2 = 5.12 \qquad s_B^2 = 5.17$
   $s_p^2 = 5.14 \qquad s_{\bar{d}} = 1.01 \qquad t = 0.059 \qquad P > 0.25 \qquad df = 18$
3. $$n_A = \frac{2(5.12)^2(1.64 + 1.64)^2}{1^2} = 567$$
   $$n_B = \frac{2(5.17)^2(1.64 + 1.64)^2}{1^2} = 578$$
4. $F = \dfrac{5.17}{5.12} = 1.01$ with 9, 9 $df \qquad P > 0.10$

5

ONE-WAY ANALYSIS OF VARIANCE

Whether or not it was apparent, the tests of comparison of samples that have been discussed to this point have had some substantial limitations. For one, they do not permit comparisons of more than two samples (or populations) at any one time although one can easily imagine instances where simultaneous comparisons of many samples would be useful. For another, it has always been implicitly assumed that the samples to be compared were essentially similar in all respects except for at most a single variable, or factor (sex, age, weight, treatment, and so forth) that was of interest. In reality this assumption has been shown to be useful but of only somewhat limited application in biological experiments since one can imagine an almost unlimited number of factors (metabolic rate, body surface area, resistance to common diseases, and so forth) varying simultaneously that could result in real variability among individuals of a population. Consequently, any one observation will undoubtedly reflect the influence of many factors: some recognized and controlled, some beyond control even if recognized, and the majority not even recognized. What is needed is a test that will not only permit comparisons of many samples simultaneously, but will also measure the relative effects of both controlled and uncontrolled, recognized and unrecognized factors on the character(s) being analyzed. The *analysis of variance* (henceforth abbreviated *anova*) is just such a test. In effect, it permits one to partition the various sources of variability inherent in any sample into at least two components: One, that due to the real effect of the recognized factors, or treatments, and two, that due to random (sampling) error which includes all variability contributed by uncontrolled factors.

One-factor design with equal replication

This simplest form of the anova is used when the observations are to be evaluated on the basis of variability attributable to but a *single recognized factor*. For example, nutrition may be the factor of interest, and so each of several groups or samples of individuals may receive a different treatment within the factor and be scored on the basis of how a treatment may affect, say, body weight (see Table 5-1). Here there are four treatment groups within the single criterion of classification: diet.

Table 5-1 Body weight (gm) in four groups of rats, each fed a different diet

	Diet				
	A	*B*	*C*	*D*	Total
	716	634	668	728	
	761	676	723	786	
	684	678	710	754	
	792	682	762	789	
	838	726	744	836	
	774	694	732	842	
$\sum_j X_{ij}$	4,565	4,090	4,339	4,735	$17{,}729 = X_{..}$
$\sum_j X_{ij}^2$	3,488,217	2,792,452	3,143,057	3,746,677	13,170,403
r	6	6	6	6	24
$\bar{X}_{i.}$	760.8	681.7	723.2	789.2	

Total variation among all observations (X_i's) can be partitioned into portions attributable to different sources of variability. One source *error*, is that variation which is due to random sampling error within each treatment and includes all uncontrolled and/or unrecognized sources of variability. The other source, *treatment*, is attributable to the effect of the variable under consideration and will equal the error term if the null hypothesis of no difference between treatments (e.g., levels of diet in this table) is true.

If we represent each of the means of the populations from which the four diets were drawn as μ_i, we may write $H_0 : \mu_1 = \mu_2 = \mu_3 = \mu_4$.

If the null hypothesis is incorrect and real differences among individuals receiving different treatments do exist, then treatment variability will ideally and most likely be significantly larger than that "background" variability due to all uncontrolled factors. We assume that each individual in a sample will randomly represent all possible uncontrolled and/or unrecognized factors and large deviations in body weight due to unusual combinations of these factors will tend to be canceled out, in the long run. The alternative hypothesis, H_1, becomes $\mu_1 \neq \mu_2 \neq \mu_3 \neq \mu_4$.

In Table 5-1, X_{ij} = the jth observation in the ith treatment, where $i = 1, 2, 3, \ldots, t$ and $j = 1, 2, 3, \ldots, r$. A dot in a subscript denotes the sum of the missing subscript, so that $X_{1.}$ is the sum of all the j observations in the ith treatment and $X_{..}^2$ is the sum of all the j observations in the i treatments, squared. (Note that the instructions in the *subscripts* are always to be completed before following those in the *superscripts*.) In this experiment there are 4 treatments, so that $t = 4$ with the number of observations (*replicates*) per treatment, r, equal to 6.

The partitioning of total variability can best be understood by first examining the *definition* formulas for total, treatment, and error sums of squares.

$$\text{Treatment } SS = r \sum_i (\bar{X}_{i.} - \bar{X}_{..})^2$$

Note that each treatment mean, $\bar{X}_{i\cdot}$, is measured as a deviation from the grand mean of all observations in all samples, $\bar{X}_{..}$, squared. As this is a measure of the variability between the i treatments or groups and the grand mean, the treatment sum of squares is often called the *between group* sum of squares.

The error sum of squares is defined as

$$\text{Error } SS = \sum_{i,j} (X_{ij} - \bar{X}_{i\cdot})^2$$

where each observation, X_{ij}, is measured approximately as a deviation within a treatment; that is, each observation in a treatment is compared to its treatment mean, $\bar{X}_{i\cdot}$, squared. The error sum of squares is, therefore, also called the *within group* sum of squares.

Figure 5-1 is a graphic representation of the rationale underlying the one-way anova. Part (a) denotes the between-group variability as measured by the treatment sums of squares. Note that each of the n

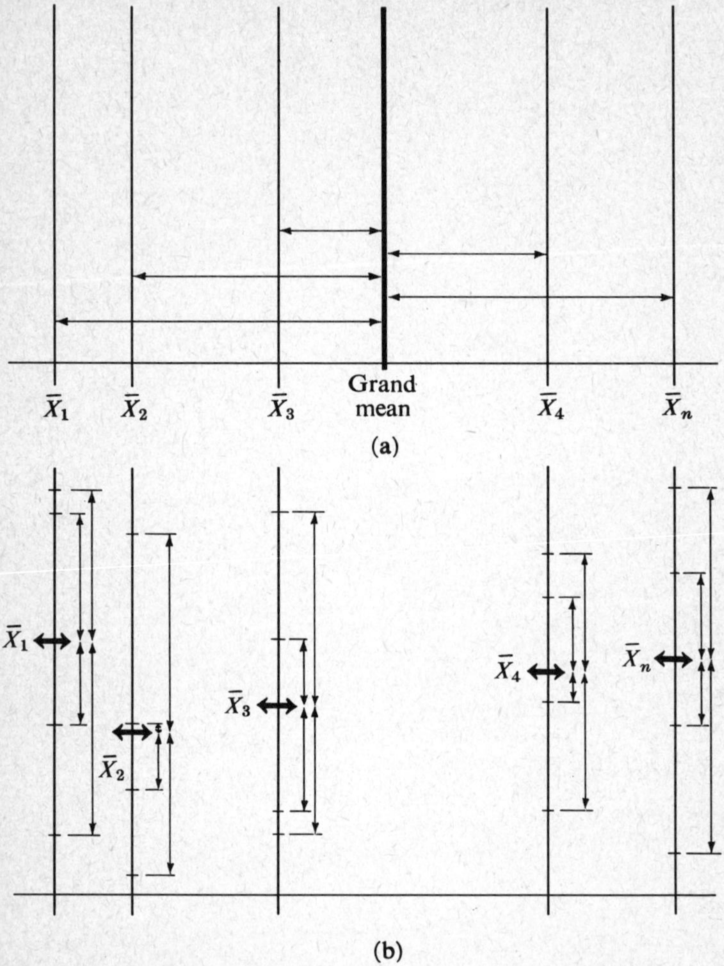

Figure 5-1 A visual representation of the treatment and error components of the anova. (a) denotes all treatment variability and (b) denotes that attributable to error.

treatment means (each is represented by a column) is measured relative to a grand mean: the measure of central tendency among the treatments. The treatment variability is measured as the sum of all deviations between all treatments and this mean:

$$r \sum_i (\bar{X}_{i.} - \bar{X}_{..})^2$$

Part (b), on the other hand, denotes the within-group (that is, error) variability. Note that the variability among the observations of a treatment is measured relative to the mean of that treatment and that the error term is the sum of all within-group variation:

$$\sum_{i,j} (X_{ij} - \bar{X}_{i.})^2$$

The between-group term includes variability due to variation among the individual observations of a treatment as well as between treatments. Thus, if the between-group term is not significantly larger than the within-group term, then it is logical to judge that the classification by treatment produces no more variability than that present among the observations themselves and, therefore, $\mu_1 = \mu_2 = \cdots = \mu_n$.

Total sum of squares logically measures total variation from all sources (both controlled and uncontrolled) and is the deviation of each observation in each treatment from the grand mean, squared:

$$\text{Total } SS = \sum_{i,j} (X_{ij} - \bar{X}_{..})^2$$

The *working formulas* differ from the definition type only in form; the former type are more readily handled on the typical desk calculator (see Table 5-2). Using the data of Table 5-1 the first step in the anova is to obtain the *correction term*, *CT*, which is analogous to the correction term described on page 22.

Table 5-2 Formulas for the one-way anova with equal replication

Source of variation	*df*	*SS*	*MS*	*F*
Treatment	$t - 1$	$\sum_i \frac{X_{i.}^2}{r} - CT$	SS_T/df	MS_T/MS_E
Error	$t(r - 1)$	$\sum_i \left(\sum_j X_{ij}^2 - \frac{X_{i.}^2}{r}\right)$	SS_E/df	
Total	$rt - 1$	$\sum_{i,j} X_{ij}^2 - CT$		

$$CT = \frac{X_{..}^2}{rt} = \frac{(\sum_{i,j} X_{ij})^2}{rt}$$

$$= \frac{(17{,}729)^2}{6(4)} = 13{,}096{,}226.7$$

Initially, one can assess variability from all sources by means of

$$\text{Total } SS = \sum_{i,j} X_{ij}^2 - CT = (716^2 + 761^2 + \cdots + 842^2) - CT$$

$$= 13{,}170{,}403 - CT = 74{,}176.3$$

From the total sum of squares one can now remove a treatment sum of squares which measures the amount of variation in body weight that can be attributed to the various treatments (diets) used in the experiment,

$$\text{Treatment } SS = \sum_i \frac{X_{i.}^2}{r} - CT = \frac{(4{,}565)^2 + \cdots + (4{,}735)^2}{6} - CT$$

$$= 13{,}135{,}744.5 - CT = 39{,}517.8$$

as well as an error sum of squares which measures variability among individuals treated alike; that is, individuals fed the same diet:

$$\text{Error } SS = \sum_i \left(\sum_j X_{ij}^2 - \frac{X_{i.}^2}{r} \right)$$

$$= 3{,}488{,}217 - \frac{(4{,}565)^2}{6} + \cdots + 3{,}746{,}677 - \frac{(4{,}735)^2}{6}$$

$$= 34{,}658.5$$

The error sum of squares may instead be obtained by subtracting the treatment SS from the total sum of squares since sums of squares are additive. In this example

$$\text{Error } SS = \text{Total } SS - \text{Treatment } SS$$

$$= 74{,}167.3 - 39{,}517.8 = 34{,}658.5$$

In that the formula for computing the error sum of squares for these data is so relatively easy, the subtraction method is not a prudent procedure here, since a mistake in calculating either total or treatment sum of squares would most probably escape detection and would result in an incorrect estimate of error variability.

A mean square, MS, is obtained by dividing the sum of squares by degrees of freedom. The mean square is an estimate of variance, since, by definition

$$s^2 = SS/df = MS$$

The error mean square, s^2 or MS_E, is an estimate of σ^2, the variance common to all observations receiving a given treatment, provided the null hypothesis of no differences in the effects of the treatments is true. If H_0 is true and all treatments come from a common population, then the only variability between treatments must be random, so under this condition, treatment mean square is an independent estimate of σ^2. Likewise, with no difference due to treatment, the treatment and error mean squares must be equal; that is, $MS_T = MS_E$. Taken another way, error mean square measures variability between individuals, but these same individuals are the ones used to measure variation between treatments. Logically, then, treatment variability must contain components of error variability. However, treatment variability contains an additional component not contained in the error term: that due to the effect of treatment.

If there is a treatment effect, then $MS_T > MS_E$. The test for the detection of a treatment effect must be a test of the equality (or inequality) of the two mean squares. As we saw earlier, one can test for equality of variances by the F ratio, which in this case is

$$F = MS_T/MS_E$$
$$= 7.6$$

with degrees of freedom $= t - 1, t(r - 1) = 3, 20$.

Table 5-3 presents the complete results of the anova performed on the data of Table 5-1. The F value, 7.6, indicates that variability among

Table 5-3 Anova for the data of Table 5-1

Source of variation	df	SS	MS	F
Treatment	3	39,517.8	13,172.6	7.6*[a]
Error	20	34,658.5	17,32.9	
Total	23	74,176.3		

[a] An asterisk after an F, t, or similar value usually denotes a significant value (i.e., $P < 0.05$).

treatments is apparently more than seven times greater than that due to all uncontrolled sources. At this point one might conclude that the variability among the treatments is due to real differences among the various diets tested. In the F table the probability of obtaining a value this large for a one-sided test by chance alone is less than .01. This implies a significant treatment effect, τ_i, (the Greek letter "tau") and the null hypothesis $H_0 : \mu_1 = \mu_2 = \mu_3 = \mu_4$ (or, if you prefer, $H_0 : \tau_1 = \tau_2 = \tau_3 = \tau_4$) is *rejected*. [An asterisk after an F, t or similar value as shown in Table 5-3 usually denotes a *significant value* (i.e., $P < 0.05$), whereas two asterisks (**) usually denote a *highly significant* value such that $P < 0.01$.]

As a general rule, a value of $F > 1$ is necessary to declare any two variances significantly different although cases where $df > 120$ in both numerator and denominator are the exception. Value of $F < 1$ can never be declared significant; the H_0: no treatment effect is always accepted in these cases. One should also remember that the F calculated in an anova is *always* one-sided since one is interested in testing only one occurrence: whether the treatment mean square is significantly greater than the error mean square. The alternative comparison (Error $MS >$ Treatment MS) implies an interest in whether the sampling errors within the treatments differ significantly from one another, which makes no sense whatever.

One-factor design with unequal replication

Often it is not possible to obtain equal numbers of replications per treatment, so that adjustments must be made in some of the calculations of the anova. However, the interpretation of the test is the same as that previously described.

$$CT = \frac{X_{..}^2}{\sum r_i}$$

$$\text{Total } SS = \sum_{i,j} X_{i.}^2 - CT$$

$$\text{Treatment } SS = \sum_i \frac{X_{i.}^2}{r_i} - CT$$

$$\text{Error } SS = \sum_i \left(\sum_j X_{ij}^2 - \frac{X_{i.}^2}{r_i} \right)$$

$$\text{Treatment } MS = \text{Treatment } SS/df$$

where degrees of freedom $= t - 1$.

$$\text{Error } MS = \text{Error } SS/df$$

where degrees of freedom $= \sum_i (r_i - 1)$, and r_i denotes the number of replicates in the ith treatment.

Tests of means

If a significant value of F (denoting a significant treatment effect) is obtained, it is then concluded that all treatments do not come from a common population or, taken another way, from populations with a common value of μ. However, it is usually an unwarranted assumption that each treatment differs from every other treatment, so that it still remains to be determined which treatments do differ significantly. There are several tests that permit comparisons of treatments—via their means—after a significant F value has been obtained. Each has its own virtues; however, we will discuss only two of them since they are reasonable representatives of the array available.

The least significant difference (lsd) test with equal replications

As you may recall, the t tests noted earlier were of the general form

$$t = \frac{\overline{X}_1 - \overline{X}_2}{s_{\bar{d}}}$$

where one had to know the values of $\overline{X}_1$, $\overline{X}_2$, and $s_{\bar{d}}$ to solve for t. The *lsd* test is based on the t test; it is intended for use with those anova in which there is some a priori reason to compare the differences only between specific pairs of treatment means. When one wishes to determine the magnitude of the difference between two means that is necessary for them to be declared significantly different at the conventional 0.05 rejection level, the formula used is

$$lsd_{.05} = t_{.05} s_{\bar{d}}$$

which is only a modification of the original t test in that now $\overline{X}_1 - \overline{X}_2$ is unknown while $t_{0.05}$ is known, as it is the appropriate tabular t value with error degrees of freedom. The pooled error term is

$$s_{\bar{d}} = \sqrt{\frac{2MS_E}{r}}$$

The $lsd_{0.05}$ value is now equivalent to *the difference between the two means which must be exceeded for the means to be declared significantly different with a probability of occurrence* ≤ 0.05.

One major advantage of this test is that but a single value need be computed for any one anova since it can be applied with equal sensitivity to all comparisons of treatment mean pairs. However, it has two of what sometimes prove to be distinct disadvantages. First, any single treatment mean can be used only once (i.e., in only one comparison) and two, for the test to have maximum reliability the treatment mean pairs intended for comparison should be chosen *before* the data are analyzed or else the experimenter may be misled by the confidence level implied by the lsd value. The reason underlying this second disadvantage is that though very tempting a comparison of the largest and smallest treatment means for no good biological reason other than that they look substantially different can in the long run often provide a significant statistic even though the means are not truly different.

Using the data of Tables 5-1 and 5-3,

$$lsd_{.05} = t_{.05}\, s_{\bar{d}} = 2.086\sqrt{\frac{2(1{,}732.9)}{6}} = 50.1 \text{ gm}$$

Due to some relationship between the various treatments, one may have earlier decided to compare $\bar{X}_C$ with $\bar{X}_D$ and $\bar{X}_A$ with $\bar{X}_B$. The sign not considered, the observed differences are

$$\bar{X}_C - \bar{X}_D = -66.0$$
$$\bar{X}_A - \bar{X}_B = 79.1$$

Since the observed differences between both pairs of treatment means are greater than 50.1, the $lsd_{0.05}$ value, we judge $X_A \neq X_B$ and $X_C \neq X_D$ with a confidence of 95 percent.

The lsd test with unequal replication

The test remains essentially the same as that for equal replication, except that the calculation of $s_{\bar{d}}$ must be modified as follows:

$$s_{\bar{d}} = \sqrt{MS_E\left(\frac{1}{r_1} + \frac{1}{r_2}\right)}$$

Duncan's multiple range test, with equal replication

Often comparisons are not planned in advance, or it may be that the comparison of one treatment mean with several others is a necessity that cannot readily be overcome using the *lsd* test. Duncan's test is more complicated and arduous than some others but it does have the distinct advantage of permitting comparisons of each treatment mean with every other. Unlike the *lsd* test, however, more than one comparison value must be computed.

For the data of Table 5-1 there is a total of 6 paired treatment mean comparisons possible: $\bar{X}_A$ and $\bar{X}_B$, $\bar{X}_A$ and $\bar{X}_C$, ..., $\bar{X}_C$ and $\bar{X}_D$ with four means ($\bar{X}_A$ through $\bar{X}_D$) in the *range* to be considered. The task at hand is to determine which, if any, means come from populations characterized by the same value of μ. The first step is to calculate $s_{\bar{x}}$, in this case a pooled standard error of the treatment means. For these data

$$s_{\bar{X}} = \sqrt{\frac{MS_E}{r}} = \sqrt{\frac{1732.9}{6}} = 16.9$$

with error mean square degrees of freedom equal to 20. *Critical values* for the ranges at the 95% and 99% levels of confidence are given in Table E. For a 95% level of confidence, for example, one enters the 0.05 row at 20 df and copies the critical values for ranges 2 through 4 since these represent the minimum and maximum numbers of means, respectively, included within any range of comparison. Then, the critical values are multiplied by $s_{\bar{x}}$ to give the *least significant range*, *lsr*, values.

Range	2	3	4
Critical values	2.95	3.10	3.18
lsr values	50.2	52.7	54.0

Next, the treatment means are ranked by size in either ascending or descending order.

Treatment	*B*	*C*	*A*	*D*
$\bar{X}$	681.6	723.2	760.8	789.2

One computes the differences between the various treatment means in the following way: largest − smallest, largest − next smallest, ..., next to smallest − smallest. Any observed difference between treatment means is considered significant if the difference is greater than the *lsr* value for that range. For example, 789.2 − 681.6 = 107.6 and the range of means encompassed (*B* to *D* when ranked by size) is four. Consequently, one compares the difference between the means to the appropriate *lsr* value of 54.0. Since 107.6 > 54.0, the lsr value, these two means are declared to be significantly different at the 0.95 level of confidence. On the other hand, $\bar{X}_C - \bar{X}_B = 41.6$. Since 41.6 < 50.2, the *lsr* value, the means of treatments *B* and *C* are not judged significantly different.

It may occasionally occur that two means are judged significantly different even though they are both contained in a larger range of means that was judged to be not significantly different. Consequently, the decision made on the basis of the larger range prevails and the two means in question must also be judged not significantly different.

In order to make the presentation of such comparisons more lucid, any two means that are not significantly different are underlined with the same line. Consequently, any two means not underscored by the same line are significantly different. Thus, for the data of Table 5-1,

$$\underline{681.6 \quad 723.2} \quad 760.8 \quad 789.2$$
$$681.6 \quad \underline{723.2 \quad 760.8} \quad 789.2$$
$$681.6 \quad 723.2 \quad \underline{760.8 \quad 789.2}$$

Duncan's test with unequal replication

As has doubtless been noted, a statistical procedure may be reasonably straightforward (and readily verifiable) when applied to data having equal numbers of replications per sample, while in those cases of unequal replications procedural complications sometimes arise. Duncan's test is no exception: it requires some relatively minor computational changes but, more important, only approximate values of the *lsr* can ever be obtained since the pooled estimate of variance, so critical to this test under this condition, may be influenced by the varying sample sizes. As most all multiple range tests suffer similarly, there is no reason to discount this test because of this weakness. However, one must always bear this shortcoming in mind when drawing final conclusions from data so analyzed.

In this procedure, one obtains a set of critical values as before but now they are not multiplied by $s_{\bar{X}}$ but by s (i.e., $\sqrt{MS_E}$) to obtain a "quasi *lsr*" value. Then, when comparing any two means this quasi *lsr* value must be multiplied by the *correction factor*

$$\sqrt{\frac{1}{2}\left(\frac{1}{r_1} + \frac{1}{r_2}\right)}$$

Anova with subsamples (equal subsample numbers within replicates)

So far we have discussed the anova with a single criterion of classification although it was possible to measure the variability attributable to two sources of variation. One source was due to the factor under consideration (treatment) and the other to differences between individuals treated alike (sampling or error). However, the anova under discussion here, while still dealing with but a single criterion of classification, permits one to measure the effect of one additional source of variation. By way of illustration, in Table 5-1 the resulting body weights were shown for each of four different diets, a diet type being assigned to each of the four groups of 6 rats each. Real variations were subsequently shown to exist in some of the diets. And, sampling variation was also seen within the groups of rats fed the same diet, but such sampling variability is always to be expected. One could ask, however, if *some* of the variability observed might really be due to nonrandomly sampled environmental factors (e.g., light, temperature, noise level, etc.). Conceivably factors such as these are not uniform within the laboratory and they possibly could affect some groups of animals more than others. A way around this problem is to use *subsamples within each treatment.* For example, one could randomly assign each of the rats to one of many cages which are, in turn, randomized as to position in the laboratory. Then, to each of the cages one of the four diets is randomly assigned, so that each treatment is represented in several locales. We will now have three potential sources of variation: One, that due to treatment; two, that among rats treated alike (i.e., within the same cage and receiving the same treatment) and, three, that among rats in different cages but receiving the same treatment. The first and second types of variation are the familiar treatment effect and sampling error, respectively. The third type, however, is new to these discussions and is called *experimental error.* It implies a flaw in the design of the experiment or the handling of the experimental units. If this last source of variability is not real, then the sampling and experimental error mean squares will be equivalent.

One need not necessarily expand an experiment in order to perform this sort of analysis. If one suspects that differences in laboratory environment may cause real variability in test results, it is a relatively simple matter to place the experimental units (rats in this example) in two or more subgroups within the same treatment category provided, of course, that the number of individuals per treatment is sufficiently large.

In the following example, a subsampling design was *simulated* by rearranging the data of Table 5-1 in the fashion shown in Tables 5-4 and 5-5. Let X_{ijk} be the weight of the kth rat in cage j receiving treatment i, with s subsamples (rats) per cage, r replications (cages) per treatment and t treatments (diets). Therefore, $k = 1, \ldots, s = 3$; $j = 1, \ldots, r = 2$; and $i = 1, \ldots, t = 4$.

$$CT = \frac{X^2_{\ldots}}{srt} = \frac{(761 + 684 + \cdots + 842)^2}{3\,(2)\,(4)} = 13{,}096{,}560.0$$

$$\text{Total } SS = \sum_{i,j,k} X^2_{ijk} - CT = (761)^2 + (684)^2 + \cdots + (842)^2 - CT$$
$$= 73{,}843.0$$

with degrees of freedom $= trs - 1 = 23$.

$$\text{Treatment } SS = \frac{\sum_i X^2_{i..}}{sr} - CT = \frac{(4{,}565)^2 + \cdots + (4{,}735)^2}{3\,(2)} - CT$$
$$= 39{,}185.2$$

with degrees of freedom $= t - 1 = 3$.

$$\text{Experimental Error } SS = \sum_i \left(\sum_j X^2_{ij.} - \frac{X_{i..}}{r} \right) \Big/ s$$
$$= \left[(2{,}237)^2 + (2{,}328)^2 - \frac{(2{,}237 + 2{,}328)^2}{2} \right.$$
$$+ \cdots + (2{,}303)^2 + (2{,}432)^2$$
$$\left. - \frac{(2{,}303 + 2{,}432)^2}{2} \right] \Big/ 3$$
$$= 4{,}953.2$$

with degrees of freedom $= t(r - 1) = 4$.

$$\text{Sampling Error } SS = \sum_{i,j,k} \left(X^2_{ijk} - \frac{X^2_{ij.}}{s} \right) = (761)^2 + \cdots + (792)^2$$
$$- \frac{(2{,}237)^2}{3} + \cdots + (754)^2 + \cdots + (842)^2$$
$$- \frac{(2{,}432)^2}{3}$$
$$= 29{,}704.9$$

with degrees of freedom $= tr(s - 1) = 16$.

Table 5-4 A regrouping of the data if Table 5-1 in order to simulate an anova with subsampling

Diet	*A*			*B*			*C*			*D*		
Cage within diet	1		2	1		2	1		2	1		2
	761		774	682		676	723		762	789		754
	684		838	634		694	668		732	728		836
	792		716	726		678	744		710	786		842
$X_{ij.}$	2,237		2,328	2,042		2,048	2,135		2,204	2,303		2,432
$X_{i..}$		4,565			4,090			4,339			4,735	
$\bar{X}_{ij.}$	745.7		776.0	680.7		682.7	711.7		734.7	767.7		810.7
$\bar{X}_{i..}$		760.8			681.7			723.2			789.2	

Table 5-5 Anova for the data of Table 5-4

Source of variation	*df*	*SS*	*MS*	*F*
Treatment	3	39,185.2	13,061.7	10.5*
Expt'l error	4	4,953.2	1,238.3	<1
Sampling error	16	29,704.9	1,856.6	
Total	23	73,843.0		

As was noted in the earliest discussions of the one-way anova, variability between individual rats of a cage [i.e., $\sum_{i,j,k}(X_{ijk} - \overline{X}_{ij.})^2$] provides an index of sampling variability, or error. However, variability between cages *within a treatment* as well as *between treatments* also contains components of sampling error since different individuals are found in each cage and each treatment. If there are *differences between cages within a treatment*, then treatment variability has a component attributable to experimental error. By elimination, then, when testing for a real treatment effect, a comparison between treatment and experimental error mean squares is logical since the treatment mean square can contain only one possible source of variability not possessed by the experimental error mean square—namely, that due to treatment effect. Therefore, H_0: no difference between treatments is obtained by

$$F = \frac{MS_T}{MS_{EE}} = 10.5$$

with $df = t - 1, t(r - 1) = 3, 4$.

Since this is a one-sided test with an observed $P < 0.025$, it is concluded that in this hypothetical example there is a significant effect due to treatment.

If real differences exist between cages treated alike, one expects $MS_{EE} > MS_{SE}$ since the former contains a source of variability (due to cage) not contained in the latter. Therefore, H_0: no difference between cages within a treatment is obtained by

$$F = \frac{MS_{EE}}{MS_{SE}} = < 1$$

with degrees of freedom $= t(r - 1), tr(s - 1) = 4, 16$.

A value of $F < 1$ can never be judged significant. Thus, in the context of these data it is concluded that no significant effect can be attributed to grouping the experimental units of a treatment into two cages.

Models

For the moment we may turn our attention to a discussion of the factor(s) used in the anova that may not affect the computations directly, but can influence the interpretation of the results. There are two basic types of models applicable to any biological problem: *fixed* and *random*.

This means that some treatments within factors are not freely chosen but are fixed by the nature of the problem, while some are randomly chosen from among an almost infinite number.

By way of example, if one wished to test the ability of a new antibiotic to kill gram-positive bacteria he might choose, say, six types of gram-positive bacteria on which to test the drug. In this example one may have perhaps chosen arbitrarily from dozens of equally relevant types of bacteria but, because of the ready availability of these six types or your knowledge of their culture, they were chosen. In this instance the treatments, or levels (bacterial types), within the factor, bacteria, are not judged to be of critical importance and so are randomly chosen; this produces an experiment of the random factor type.

If the antibiotic were developed with the express purpose of killing, say, six specific strains of a gram-positive bacterial species, logically the types used in the experiment would not be randomly chosen, but fixed. Sometimes models are *mixed* in that they involve both random and fixed treatments, but they can be handled without any real difficulty.

The basic linear additive model for the one-way anova is

$$X_{ij} = \mu + \tau_i + \varepsilon_{ij}$$

where X_{ij} is the jth observation in the ith treatment and τ_i is a treatment effect. As an equivalent alternative one may instead write

$$X_{ij} = \mu_i + \varepsilon_{ij}$$

where

$$\mu_i = \mu + \tau_i$$

It was upon the assumptions of this model that earlier in this chapter we were able to equate

$$H_0 : \mu_1 = \mu_2 = \mu_3 = \mu_4$$

to

$$H_0 : \tau_1 = \tau_2 = \tau_3 = \tau_4$$

In the fixed model, the treatments, τ_i's, are fixed by the experimenter. For the random model, the τ_i's are a random sample from a population of τ's for which the mean is zero and variance is σ_τ^2.

In the fixed model, no matter how many times the experiment were repeated, the same set of τ's (e.g., the 6 bacterial types) would be used.

For the random model, repeating the experiment would introduce new τ's (e.g., gram-positive bacterial types) but all would be drawn from the same population (all gram-positive bacterial types). Then, for the fixed model one examines a particular treatment while for the random model one examines the population of treatments as represented by the various observations, τ_i's, drawn from it.

The data of Table 5-1 clearly represent an experiment of the fixed model type since the diets were chosen only after careful consideration. As such, the true or expected value of treatment variance, estimated by MS_T, contains two components as noted earlier: the variability due to the effects of the various treatments, $r \sum \tau_i^2/t - 1$, as well as that due to sampling error, σ_ε^2. If there is no real difference among the treatments, then MS_T should equal MS_E and the correct test for the presence or absence of treatment effect is the F test as was performed earlier: MS_T/MS_E.

In a one-way anova for a random model experiment, the MS_T would also contain two components: that due to variability between treatments which arises from the sample-to-sample differences in the τ's, $r\sigma_\tau^2$, and the random error variance σ_ε^2. In this instance the appropriate test for treatment effect is also $F = MS_T/MS_E$.

In summary, while the F tests are similar in both models, the parameters to be estimated vary. For the fixed model one tries to *estimate* $\sum \tau_i^2/t - 1$ from $F = [(\sum \tau_i^2/t - 1) + \sigma_\varepsilon^2]/\sigma_\varepsilon^2$ and for the random model, σ_τ^2 from $F = (\sigma_\tau^2 + \sigma_\varepsilon^2)/\sigma_\varepsilon^2$.

PROBLEMS

1. A hematologist desired to test two new anticoagulant drugs against a drug widely and traditionally used in anticoagulant therapy though known to have undesirable side effects if continued over long periods. Thirty-seven laboratory animals of the same age and sex were randomly chosen to receive either the standard or one of the experimental drugs in a dose equilibrated to body weight (1 mg/kg). The data recorded give the whole blood clotting time (min) after 2 weeks of drug therapy.

Standard	*Experimental 1*	*Experimental 2*
20	16	22
19	21	19
21	19	25
16	19	21
19	17	21

Standard	*Experimental 1*	*Experimental 2*
14	16	24
17	19	22
19	22	19
18	20	22
15	15	23
20	18	24
20		26
		27
		21

(a) Compute the F test appropriate to testing H_0: no difference between treatments; that is, $\mu_1 = \mu_2 = \mu_s$. (b) Is the null hypothesis accepted or rejected? (c) It now remains to determine which means do vary. Though $\bar{X}_s$ and $\bar{X}_1$ are similar and $\bar{X}_2$ is the most different of the three, there is, however, no a priori reason to compare $\bar{X}_2$ to the others on the basis of the experimental design. Thus, an *lsd* test would not be prudent; instead, a Duncan's test would be more logical. On the basis of this test what do you conclude about the treatment means?

2. In each of 4 geographic locales several small rodents (all of one species) were captured and the number of ectoparasites adhering to each animal was counted and recorded as shown.

Locale:	*1*	*2*	*3*	*4*
	25	19	11	46
	15	7	10	15
	17	5	21	21
	25	10	19	11
	10		21	19
	10			

By means of an anova determine whether $\mu_1 = \mu_2 = \mu_3 = \mu_4$ at the 0.05 level of significance.

3. From each of seven widely separated geographic locales an entomologist trapped 5 of a particular species of moth and counted the number of wing spots on each individual in order to determine whether there was a significant difference between locales in regard to this character.

Locale:	*1*	*2*	*3*	*4*	*5*	*6*	*7*
	36	40	42	56	47	51	46
	37	44	47	53	43	50	41
	43	50	40	45	47	40	42
	39	50	41	56	46	42	44
	38	48	47	52	47	43	54

(a) Compute an anova on the data shown above. (b) Do you accept or reject H_0 at the 0.05 level of significance? (c) For practice compute the *lsd* value and compare $\bar{X}_4$ and $\bar{X}_1$, then $\bar{X}_2$ and $\bar{X}_3$, for significance of differences at the 0.05 level.

4. Given the following data regarding an anova with subsampling

$$\begin{aligned} SS_T &= 665.68 & df &= 2 \\ SS_{EE} &= 1720.68 & &= 9 \\ SS_{SE} &= 15.62 & &= 12 \\ SS_{TOT} &= 2401.97 & &= 23 \end{aligned}$$

determine (a) the number of treatments used within the experiment and (b) the number of experimental units within a treatment and the number within each subsample. (c) Compute the appropriate mean squares necessary for the statistical analysis. (d) Determine whether there is a significant component of sampling error and experimental error.

5. Devise hypothetical experiments in which one would invoke (a) the fixed model, (b) the random model, (c) the random model of anova.

ANSWERS TO PROBLEMS

1. (a) $SS_{TOT} = 339.57$ $SS_T = 161.93$ $SS_E = 177.64$
   $\bar{X}_s = 18.17$ $\bar{X}_1 = 18.36$ $\bar{X}_2 = 22.57$
   $F = 15.496$ (2, 34 *df*)
   (b) Rejected ($P < 0.005$)
   (c) $s = 2.29$; Correction factor $= \sqrt{\frac{1}{2}\left(\frac{1}{r_1} + \frac{1}{r_2}\right)}$
   $\mu_s = \mu_1 \neq \mu_2$
2. $SS_{TOT} = 1548.55$ $SS_T = 329.40$ $SS_E = 1219.15$
   $F = 1.44$ (3, 16 *df*) $p > 0.10$; thus H_0 accepted.
3. (a) $SS_{TOT} = 951.89$ $SS_T = 503.09$ $SS_E = 448.80$
   $F = 5.23$ (6, 28 *df*)
   (b) $P < 0.005$; thus H_0 rejected.
4. (a) 3
   (b) 4 and 2
   (c) $SS_T = 332.84$ $SS_{EE} = 191.19$ $SS_{SE} = 1.30$
   (d) $F = 147.07$ ($P < 0.01$) and $F = 1.74$ ($P > 0.05$), respectively.

6

FACTORIAL ANALYSIS OF VARIANCE

In the previous chapter, anovas were performed on data classified by a single criterion, or factor. The factorial anova is performed on data classified by two or more factors of equal rank; that is, they are afforded equal weight in the analysis. Each factor is handled as previously in that it is divided into two or more treatments or levels of the factor.

It is not sufficient, however, to know that several factors are varying in a sample in order for the factorial analysis to be applied. This type of analysis requires that one most basic condition always be fulfilled: Each level of one factor must appear with every level of the other factor(s) as indicated in Table 6-1. Though this is no slight complication when it comes to running the experiment, it pays the substantial dividends of permitting not only measurement of variability due to any or all controlled factors and levels, but also all possible interactions between them.

Table 6-1 General form of a 4 × 2 two factor anova

		Factor *A*			
	Level	a_1	a_2	a_3	a_4
Factor B	b_1	a_1b_1	a_2b_1	a_3b_1	a_4b_1
	b_2	a_1b_2	a_2b_2	a_3b_2	a_4b_2

Consider an experiment which describes the effects of two factors—*A*, drug concentration and *B*, sex—on protein added to bodies of laboratory mice given fixed caloric intakes. It may be that one has no a priori knowledge of either factor influencing protein formation although one or both factors must be at least suspect in that they were chosen for testing. The aim here is to evaluate the effects of all levels of both factors individually, and in combination. If neither sex nor drug concentration influences protein formation one would expect, *ideally*, identical values for amount of protein at every level of either factor; specifically, the cells of Table 6-1 would contain eight absolutely identical values. In reality, obviously, sampling variability and sample size would cause differences among the observed values. But, at what point does sampling variability end and the effects of the treatments—acting either individually or in concert (that is, interacting) begin? In essence, the factorial anova measures departures from equality for each of the observations at the various combinations of levels and permits one to partition variability into components attributable to chance, effect of a treatment, and combinations of treatments (interaction).

In this factorial design there will be four levels within Factor A denoted a_1, a_2, a_3, a_4 and two levels within $B: b_1$ and b_2. This arrangement comprises a slightly advanced form within the framework of the simplest type of factorial: the *two factor* design. Specifically, this example is of a 4×2 factorial since there are four and two levels within the two factors; logically, there must be $4 \times 2 = 8$ cells in the table.

The notation $a_i b_j$ can refer to the mean of observations made on that combination of treatments or simply to the combination itself: Factor A at level a_i and Factor B at level b_j. There is no restriction on the number of observations contained within a cell value although it is strongly suggested that there always be at least two observations made for every combination of factors. (There are ways of handling two factor anovas without replication but they require special circumstances and assumptions.) Likewise, differences in the numbers of observations per cell should be avoided whenever possible, as a strict adherence to the computations required under the circumstance of unequal replication can become quite laborious. Consequently, if only a few cells have fewer than the basic number of observations and they differ only slightly from the other cells, practically speaking these discrepancies would best be ignored and the data treated as though equally replicated. (In the long run they make little difference to the quality of the test but if strictly considered would require the expenditure of a considerable amount of extra computation time.)

For any size factorial anova, the basic computations are similar. For the two factor type a treatment sum of squares is calculated which is then partitioned into three components:

1. $SS(A)$: main effect of Factor A
2. $SS(B)$: main effect of Factor B
3. $SS(AB)$: interaction of Factors A and B

Before entering into the computations of this anova it would be wise to at least quickly scan the data of the experiment to be described as shown in Table 6-2. The format is like that of Table 6-1. The data are

Table 6-2 Effect of drug concentration (Factor A) and sex (Factor B) on amount protein (gm) added to the body weights of mice given a fixed caloric intake [a]

		Factor A			
	Level	a_1	a_2	a_3	a_4
Factor B	b_1	4.9	3.6	3.0	2.0
		6.1	5.9	2.7	1.5
		5.1	5.7	3.8	2.2
	$X_{i1.}$	16.0	15.2	9.5	5.7
	$\sum X_{i1k}^2$	87.2	80.3	30.7	11.1
	$\bar{X}_{i1.}$	5.4	5.1	3.2	1.9
	b_2	8.5	8.0	5.9	3.9
		10.0	9.4	6.2	4.2
		10.6	8.4	5.9	4.3
	$X_{i2.}$	29.1	25.8	18.0	12.4
	$\sum X_{i2k}^2$	284.6	222.9	108.1	51.3
	$\bar{X}_{i2.}$	9.7	8.6	6.0	4.1

Treatment Totals

Level	a_1	a_2	a_3	a_4	Total
b_1	16.0	15.2	9.5	5.7	46.4
b_2	29.1	25.8	18.0	12.4	85.3
Total	45.1	41.0	27.5	18.1	131.7

[a] Concentration of drug increases from a_1 to a_4; males $= b_1$, females $= b_2$. Each observation was made on a different animal.

in a sense condensed in the lower (Treatment Total) portion of the table with each of the four a_i *columns* being seen to represent the effects of a particular level of drug concentration at both levels of B, sex. For example, combinations a_1b_1 and a_1b_2 have level a_1 in common, but the levels of B vary. It can be seen by casual inspection that there is a *relatively* large difference between the b_j's, holding a_1 constant (16.0 versus 29.1), suggesting that the two B levels may each produce a rather different response in the experimental units at level a_1. Similar comparisons of a_2b_1 and $a_2b_2, \ldots, a_4b_1$ and a_4b_2 provide about the same picture: The responses at the two levels of B vary to a noticeable degree within each level of A, which may eventually prove to indicate a significant *main effect of Factor B*. When examining the observations in the two *rows*, b_1 and b_2, one can now make the analogous comparisons: the effects of the A levels within either level of B. In this type of comparison one can note a gradual but overall marked decrease in protein formed when going from a_1 through a_4 in both levels of B. This may eventually prove indicative of a significant *main effect of Factor A*.

In a slightly different fashion one can make a preliminary—but by no means conclusive—examination for the *interaction of A and B* in the following way: If a response to A, that is, the decrease in protein formed from, say, a_1 to a_2 is substantially greater for b_2 than for b_1 (16.0 to 15.2 at b_1; 29.1 to 25.8 at b_2), then a unique interaction of the factors may be present. In these data, though, since the magnitude of the differences between 16.0 and 15.2, and 29.1 and 25.8 as examples are not grossly different, this assumption does not seem to be warranted at this time. Of course, this entire discussion has thus far been based on subjective evaluations and can in no ways bypass the actual computations of the anova.

Let a, b, and r denote the number of levels of factors A and B and the number of observations (replicates) per cell, respectively; there are then, $4 \times 2 = 8$ treatment combinations. The computations to be described below apply equally well to all factorial anovas regardless of the numbers of factors and levels. Thus, in this example $i = 1, \ldots, a = 4; j = 1, \ldots, b = 2; k = 1, \ldots, r = 3$; so that X_{ijk} denotes the kth observation at the jth level of B and the ith level of A. Then

$$CT = \frac{X_{\cdots}^2}{abr} = \frac{(16.0 + 15.2 + \ldots + 12.4)^2}{4\,(2)\,(3)} = \frac{(131.7)^2}{24} = 722.7$$

$$\text{Total } SS = \sum_{i,j,k} X_{ijk}^2 - CT = (4.9)^2 + (6.1)^2 + \cdots + (4.3)^2 - CT$$
$$= 153.5$$

$$\text{Treatment } SS = \frac{\sum_{i,j} (X_{ij.})^2}{r} - CT$$
$$= \frac{(16.0)^2 + (15.2)^2 + \cdots + (12.4)^2}{3} - CT$$
$$= 143.9$$

The treatment sum of squares may now be partitioned into the effects attributable to the factors:

$$SS(A) = \frac{\sum_i (X_{i..})^2}{rb} - CT = \frac{(45.1)^2 + (41.0)^2 + \cdots + (18.1)^2}{3\ (2)} - CT$$
$$= 77.1$$

and

$$SS(B) = \frac{\sum_j (X_{.j.})^2}{ra} - CT = \frac{(46.4)^2 + (85.3)^2}{3\ (4)} - CT = 63.1$$

and the interaction of these factors

$$SS(AB) = \text{Treatment } SS - SS(A) - SS(B) = 3.7$$

Although it is almost never advisable to compute a sum of squares by subtraction for reasons noted earlier, in the case of factorials more advanced than the 2 × 2 type the computation of the interaction sum of squares is quite laborious.

One can compute that portion of total variability that cannot be assigned to treatments as

$$\text{Error } SS = \sum_{i,j} \left[X_{ijk}^2 - \frac{(X_{ij.})^2}{r} \right] = 87.2 - \frac{(16.0)^2}{3}$$
$$+ \cdots + 51.3 - \frac{(12.4)^2}{3} = 9.6$$

Table 6-3 presents the results of this anova. The F's are calculated as

$$\text{Effect of Factor } A = \frac{MS_A}{MS_E}$$

$$\text{Effect of Factor } B = \frac{MS_B}{MS_E}$$

$$\text{Interaction } (AB) = \frac{MS_{AB}}{MS_E}$$

Table 6-3 Anova for data of Table 6-2

Source of variation	*df*	*SS*	*MS*	*F*
Treatments	$ab - 1 = 7$	143.9	20.6	
(A)	$a - 1 = 3$	77.1	25.7	42.8**
(B)	$b - 1 = 1$	63.1	63.1	105.2**
(AB)	$(a - 1)(b - 1) = 3$	3.8	1.3	2.2 ns
Error	$ab(r - 1) = 16$	9.6	0.6	
Total	$rab - 1 = 23$	153.5		

From the significant F values calculated for both factors A and B ($P < 0.005$ in both cases) there is a strong indication that both influence protein formation. However, since the interaction term is not judged significant ($P > 0.10$) it appears likely that these two factors operate independently, i.e., without interaction, since *no particular combinations of levels* (a_i with b_j) *produce unique or unexpected results.*

The F values were obtained with the use of a single error term. In evaluating the F's it is implicitly assumed that the observations in each cell have come from populations with a common variance, regardless of level or combination of levels. Sometimes this assumption is unwarranted. If the variability between observations within the cells changes to an obvious degree from cell to cell, then one might well be skeptical of the validity of the error term (which may be meaningless since it is meant as an estimate of the variance common to all cells) and, more especially, the validity of an apparently significant interaction effect since heterogeneity of the cell variances will generally cause the interaction sum of squares to be an overestimate of the true value. Very little can be done to remedy such an occurrence short of ignoring the problem if the variability is not deemed "overly" large or, at the other extreme, disregarding the data completely.

It is also possible to correctly calculate and accept a significant interaction effect and yet not obtain a significant main effect for one or all of the factors involved in the interaction. Such observations logically are hard to explain. For example, it may be that the levels of the factor judged not significant are really important but balance each other out so neatly from cell to cell that no main effect appears. Or, it may be that there really is no difference between levels of any factor if in each cell there is as much difference from, say, concentration to concentration as there is from male to female. The first alternative requires a fortuitous set of circumstances while the second would dictate that no real interaction could be present. In either case, such an occurrence requires that the individual observations deserve a great deal of scrutiny. For the data of Table 6-2 the interaction of A with B was not judged significant. However, assuming *only for the moment* that it were significant, the interaction could be subjected to further scrutiny by an examination of the *simple effects*. In this case one would begin by setting up a table of the form shown as Table 6-4, where each of the cell values is the *sum*

Table 6-4 An examination of the simple effects[a]

	b_1	b_2	$b_2 - b_1$
a_1	$a_1 b_1$	$a_1 b_2$	$a_1 b_2 - a_1 b_1$
a_2	$a_2 b_1$	$a_2 b_2$	$a_2 b_2 - a_2 b_1$
a_3	$a_3 b_1$	$a_3 b_2$	$a_3 b_2 - a_3 b_1$
a_4	$a_4 b_1$	$a_4 b_2$	$a_4 b_2 - a_4 b_1$

[a] Each $a_i b_j$ here represents a sum over all replicates in the cell.

over all replicates within that combination of levels. The values $b_2 - b_1$ are the simple effects since, by definition, they are *the net differences between any levels of one factor within a level of another factor*. If one of the simple effects is obviously larger than the others, then one has an a priori reason for suspecting that this particular combination of levels is of prime importance in producing the significant AB interaction. This conclusion can be verified, or refuted, by calculating the *sums of squares for the simple effects* as follows:

$$SS\,(B \text{ within } a_1) = \frac{[(a_1b_2) - (a_1b_1)]^2}{ra}$$

and, in general,

$$SS\,(B \text{ within } a_i) = \frac{[(a_ib_2) - (a_ib_1)]^2}{ra}$$

with $a - 1$ degrees of freedom. Mean squares are computed for the *four* such sums of squares, as are F tests of the form

$$F = \frac{MS\,(B \text{ within } a_i)}{MS_E}$$

Any significant F value(s) should provide the experimenter with a strong suggestion of those level combinations that do interact.

Models in the factorial

The data of Table 6-2 are obviously from a *fixed* model experiment and, as can be deduced from Table 6-5, the F tests used in this anova were entirely adequate to provide estimates of the real variances attributable to the various sources of variability since dividing either main effect or interaction mean square by the error mean square provides an estimate of the true variance attributable to that component. (Here, α and β refer to treatments within Factors A and B, respectively.)

Table 6-5 Expected values of mean squares in a two factor anova

Source	Fixed model	Random model
A	$\sigma_\varepsilon^2 + rb\sum\alpha^2/a-1$	$\sigma_\varepsilon^2 + r\sigma_{AB}^2 + ra\sigma_A^2$
B	$\sigma_\varepsilon^2 + ra\sum\beta^2/b-1$	$\sigma_\varepsilon^2 + r\sigma_{AB}^2 + rb\sigma_B^2$
AB	$\sigma_\varepsilon^2 + r\sum(\alpha\beta)^2/(a-1)(b-1)$	$\sigma_\varepsilon^2 + r\sigma_{AB}^2$
Error	σ_ε^2	σ_ε^2

On the other hand, an equivalent *random* model experiment requires that different ratios of variances be considered since by inspection of this table it is clear that no single denominator variance would provide us with estimates of the real variances. Then, for the random model

$$F_A = \frac{MS_A}{MS_{AB}}$$

$$F_B = \frac{MS_B}{MS_{AB}}$$

$$F_{AB} = \frac{MS_{AB}}{MS_E}$$

Mixed model experiments are handled in the same manner as the random model type: F ratios are computed between mean squares which differ only in the component being tested.

Three factor tests

Tests of this type are considerably more complicated than the two factor type but are computed in the same general manner. For example, X_{ijkl} represents the lth observation in the kth level of Factor C, jth level of B, and ith level of A. Let a, b, c, and r denote the numbers of levels of Factors A, B, and C and the number of replications per cell, respectively, so that there is a total of abc treatment combinations. Thus, $i = 1, \ldots, a$; $j = 1, \ldots, b$; $k = 1, \ldots, c$; $l = 1, \ldots, r$. Examine Table 6-6 as one possible type of three factor anova.

In this type of test the treatment sum of squares can now be partitioned into seven components: three main effects [$SS(A)$, $SS(B)$, $SS(C)$] three *first order* interactions [$SS(AB)$, $SS(AC)$, $SS(BC)$] and one *second order* interaction, $SS(ABC)$.

$$CT = \frac{X^2_{....}}{abcr}$$

$$\text{Total } SS = \sum_{i,j,k,l} X^2_{ijkl} - CT$$

$$\text{Error } SS = \sum_{i,j,k} \left[X^2_{ijkl} - \frac{(X_{ijk.})^2}{r} \right]$$

$$\text{Treatment } SS = \sum_{i,j,k} \frac{(X_{ijk.})^2}{r} - CT$$

The treatment sum of squares can be partitioned as indicated above. To indicate the general patterns of computation one of each of the representative types is shown below with degrees of freedom shown in Table 6-7.

$$SS(A) \quad = \sum_{i} \frac{(X_{i...})^2}{rbc} - CT$$

$$SS(AB) \quad = \sum_{i,j} \frac{(X_{ij..})^2}{rc} - CT - SS(A) - SS(B)$$

$$SS(ABC) = \sum_{i,j,k} \frac{(X_{ijk.})^2}{r} - CT - SS(A) - SS(B) - SS(C)$$
$$- SS(AB) - SS(AC) - SS(BC)$$

Table 6-6 A possible format for a three factor, $4 \times 2 \times 3$ factorial anova [a]

		Factor C		
Factor A	Factor B	c_1	c_2	c_3
a_1	b_1	$a_1b_1c_1$	$a_1b_1c_2$	$a_1b_1c_3$
	b_2	$a_1b_2c_1$	$a_1b_2c_2$	$a_1b_2c_3$
a_2	b_1	.	.	.
	b_2	.	.	.
a_3	b_1	.	.	.
	b_2	.	.	.
a_4	b_1	.	.	.
	b_2	$a_4b_2c_1$	$a_4b_2c_2$	$a_4b_2c_3$

[a] Each treatment combination, $a_i b_j c_k$, represents the sum of r replicates.

Assuming a *fixed* effects model, the F tests are computed in the usual manner:

$$F = \frac{MS_T}{MS_E}$$

where $MS_T = MS_A, MS_B, \ldots, MS_{ABC}$. If a significant AB, AC, or BC (first order) interaction is obtained, it is interpreted in the manner discussed in the two factor anova. However, if a significant second order interaction (ABC) is indicated, it is generally not immediately

Table 6-7 Degrees of freedom for the three factor anova

Source	*df*
A	$a-1$
B	$b-1$
C	$c-1$
AB	$(a-1)(b-1)$
AC	$(a-1)(c-1)$
BC	$(b-1)(c-1)$
ABC	$(a-1)(b-1)(c-1)$
Error	$(r-1)(abc-1)$
Total	$abcr-1$

apparent as to how it should be interpreted, especially if all the first order interactions are also judged significant. For example, under these circumstances it can be viewed either as an interaction of AB with C (that is, the AB interaction is not uniform for all levels of C); and, similarly AC with B or BC with A. One might begin the analysis of the second order interaction by first examining the first order interactions. Usually the simplest approach is the best; since there are only two levels of B and three of C, one may first examine the BC interaction by setting up a format like that shown in Table 6-8. The effect(s) of Factor A are removed from the computations by the simple expedient of using *total* values of a_i in each of the cells so that the value b_1c_1, for example, is really $\sum_i a_i b_1 c_1 = a_1 b_1 c_1 + \cdots + a_4 b_1 c_1$, and so on. The values $b_2 - b_1$ are the simple effects since they are the net differences between levels of one factor within a level of another factor.

If one of the $b_2 - b_1$ differences is obviously larger than the others, then one has grounds for immediately suspecting that this combination of levels, $b_j c_k$, is of prime importance in producing the significant BC

Table 6-8 The simple effects of Factor B within Factor C[a]

	b_1	b_2	$b_2 - b_1$
c_1	$b_1 c_1$	$b_2 c_1$	$b_2 c_1 - b_1 c_1$
c_2	$b_1 c_2$	$b_2 c_2$	$b_2 c_2 - b_1 c_2$
c_3	$b_1 c_3$	$b_2 c_3$	$b_2 c_3 - b_1 c_3$

[a] The effects of Factor A are held constant.

Table 6-9 Expected values of mean squares in a three factor anova

Source	Fixed model	Random model
A	$\sigma_\varepsilon^2 + rbc\Sigma\alpha^2/a-1$	$\sigma_\varepsilon^2 + r\sigma_{ABC}^2 + rc\sigma_{AB}^2 + rb\sigma_{AC}^2 + rbc\sigma_A^2$
B	$\sigma_\varepsilon^2 + rac\Sigma\beta^2/b-1$	$\sigma_\varepsilon^2 + r\sigma_{ABC}^2 + rc\sigma_{AB}^2 + ra\sigma_{BC}^2 + rac\sigma_B^2$
C	$\sigma_\varepsilon^2 + rab\Sigma\gamma^2/c-1$	$\sigma_\varepsilon^2 + r\sigma_{ABC}^2 + rb\sigma_{AC}^2 + ra\sigma_{BC}^2 + rab\sigma_C^2$
AB	$\sigma_\varepsilon^2 + rc\Sigma(\alpha\beta)^2/(a-1)(b-1)$	$\sigma_\varepsilon^2 + r\sigma_{ABC}^2 + rc\sigma_{AB}^2$
AC	$\sigma_\varepsilon^2 + rb\Sigma(\alpha\gamma)^2/(a-1)(c-1)$	$\sigma_\varepsilon^2 + r\sigma_{ABC}^2 + rb\sigma_{AC}^2$
BC	$\sigma_\varepsilon^2 + ra\Sigma(\beta\gamma)^2/(b-1)(c-1)$	$\sigma_\varepsilon^2 + r\sigma_{ABC}^2 + ra\sigma_{BC}^2$
ABC	$\sigma_\varepsilon^2 + r\Sigma(\alpha\beta\gamma)^2/(a-1)(b-1)(c-1)$	$\sigma_\varepsilon^2 + r\sigma_{ABC}^2$
Error	σ_ε^2	σ_ε^2

interaction. The validity of this conclusion can be checked by calculating the sums of squares for the simple effects of B as follows:

$$SS\,(B \text{ within } c_1) = \frac{[(b_2 c_1) - (b_1 c_1)]^2}{2ra}$$

and, in general

$$SS\,(B \text{ within } c_k) = \frac{[(b_2 c_k) - (b_1 c_k)]^2}{2ra}$$

with $b - 1$ degrees of freedom.

Of course, this type of analysis can be extended to any pair of factors using the analogous computational procedure.

Mean squares are computed for all such sums of squares, as are F tests of the form

$$F = \frac{MS\,(B \text{ within } c_k)}{MS_E}$$

When all indicated comparisons are made, one should have a reasonably good idea as to which is (are) the particular level(s) within the factors that interact to produce the ABC interaction.

Had the data been from a random or mixed model, the F tests within the initial anova would have to be computed on the basis of decision made from an examination of Table 6-9; that is, on the basis of mean squares which *differ only in the component being tested.* (Here, α, β, and γ refer to treatments within factors A, B, and C, respectively.)

PROBLEMS

1. The following data record weight gain (gm) among 12 rats of comparable age over a three-month period. Note that the data are classified by diet type and sex.

	Diet (A)	
Sex (B)	a_1	a_2
	679	476
b_1 (male)	709	538
	699	592
	594	508
b_2 (female)	677	539
	657	505

(a) Compute the appropriate factorial anova. (b) Is there a significant effect of Factor A or Factor B? (c) What is the nature of the interaction between Factors A and B? (d) What does the error term measure in such an experiment? (e) Is there any reason to partition the effects of B within a_i; A within b_i? Why?

2. The following data represent a three factor anova where each value represents the *sum* of 5 replicates.

Factor A	Factor B	Factor C				
		c_1	c_2	c_3	c_4	c_5
	b_1	246	127	97	149	164
a_1	b_2	201	130	99	131	170
	b_3	271	136	102	138	150
	b_1	207	87	60	86	111
a_2	b_2	124	91	72	81	104
	b_3	158	89	74	99	117
	b_1	129	52	46	62	93
a_3	b_2	79	51	52	50	54
	b_3	142	51	32	51	63

Assuming a fixed effects model, compute the complete anova and display it in the conventional tabular form. Because the original observations have been condensed in the table (i.e., each cell represents the sum of 5 observations), it is impossible to estimate the within subgroup (i.e., Error) sum of squares. Hence, if one makes the assumption that $A \times B \times C$ interaction is zero, the term MS_{ABC} will estimate error variance so that all F's may be computed with this quantity as the denominator.

3. The concentration of a particular blood clotting constituent was measured (mg/100 ml) in 44 adult males and females of two closely related species of dogs.

	Factor A (sex)			
	a_1 (male)		a_2 (female)	
Factor B (Species)	b_1	b_2	b_1	b_2
	1.29	1.29	1.26	1.26
	1.21	1.37	1.31	1.29
	1.22	1.36	1.21	1.22
	1.25	1.33	1.27	1.21
	1.28	1.32	1.28	1.26
	1.24	1.31	1.25	1.25
	1.31	1.34	1.24	1.24
	1.31	1.39	1.22	1.27
	1.26	1.34	1.32	1.24
	1.27	1.30	1.31	1.31
	1.30	1.30	1.28	1.20

(a) Compute the anova for this 2 × 2 factorial. (b) By means of F tests determine whether the effects of species, sex, or both influence this constituent of blood. (c) Assuming for the moment that the results of the anova were not available, compute a Duncan's test to determine if $H_0: \mu_1 = \mu_2 = \mu_3 = \mu_4$ is valid at the 0.05 level of significance.

ANSWERS TO PROBLEMS

1. (a) $SS(A) = 61{,}204.08$ $SS(B) = 3{,}780.75$ $SS(AB) = 918.75$
   $SS_E = 11{,}666.70$ $SS_{TOT} = 77{,}570.25$
   (b) $F_A = 41.97$ $(P < 0.005)$ $F_B = 2.59$ $(P > 0.05)$
   (c) $F_{AB} < 1$
   (e) None, since the interaction term is not judged significant.
2. $SS(A) = 57{,}116.1$ $SS(B) = 3{,}758.8$ $SS(C) = 55{,}545.5$
   $MS(A) = 28{,}558.0$ $MS(B) = 1{,}879.4$ $MS(C) = 13{,}886.4$
   $SS(AB) = 97.1$ $SS(AC) = 3{,}685.9$ $SS(BC) = 5{,}264.5$
   $MS(AB) = 24.3$ $MS(AC) = 460.7$ $MS(BC) = 658.1$
   $SS(ABC) = 1{,}034.9$ $MS(ABC) = 64.7$
   $F_A = 441.39$ $(P < 0.005)$ $F_B = 214.63$ $(P < 0.005)$ $F_C = 29.05$ $(P < 0.005)$
   $F_{AB} < 1$ $F_{AC} = 7.12$ $(P < 0.005)$ $F_{BC} = 10.17$ $(P < 0.005)$
3. (a) $SS_{TOT} = 0.893$ $SS(A) = 0.0180$ $SS(B) = 0.0059$
   $SS(AB) = 0.0188$ $SS_E = 0.0465$
   (b) $F_A = 3.913$ $(P > 0.05)$ $F_B = 1.293$ $(P > 0.10)$
   $F_{AB} = 4.087$ $(P > 0.05)$
   (c) $\mu_1 = \mu_3 = \mu_4 \neq \mu_2$

7

LINEAR REGRESSION AND CORRELATION ANALYSES

A *regression analysis* of one variable upon another is a means of estimating just how much of an increase or decrease in one variable may be expected, per unit of increase or decrease in the other. One variable, then, changes as a function of the other, so that we will consider one variable *independent* (usually denoted X) and the other *dependent* (denoted Y) upon the first.

By way of example, during his first fourteen or so years, a boy's height, Y, is in some fashion clearly dependent upon his age, X. If the relationship between these two variables is precisely known, it is clear that one can predict the height of any boy of a given age, or taken another way, how a given boy's height will change per unit of age. The whole purpose of the regression analysis, then, is to quantify the nature of the relationship between the variables. Prediction of the value of the dependent variable—given the value of the independent—is obviously of great importance, but regression analyses are also useful in determining mechanisms underlying various biological processes. If you demonstrate by regression analysis that, for example, the degree of affliction with a certain disease is related to the concentration of a glandular product in the blood, you are provided with a strong clue as to the particular gland controlling the disease.

Calculation of a linear regression

Graphing the values of the two variables is somewhat akin to a regression analysis. If one plots each combination of variables (X_i, Y_i) as a point having corresponding X and Y coordinates, a scatter diagram will result; it will follow some sort of straight or curved line relationship if there is, indeed, a relationship between the variables. If there is none, the points will simply lie scattered randomly on the paper. If the points follow a straight line, the relationship is *linear*; if it is a curved line that best describes the pattern of points, the relationship is *curvilinear*. This chapter will consider only the former type of relationship although in Chapter 9 we will discuss the latter. The graphic representation of the linear regression is the fitting of a straight line to the data; not just any straight line but *the line of best fit*, which almost always precludes fitting the line by inspection and requires instead a precise mathematical (that is, regression) procedure.

The equation for any point on a straight line may be expressed mathematically in the form:

$$Y_i = a + bX_i$$

where X_i = value of the independent variable,
Y_i = value of the dependent variable,
a = Y-intercept (the point where the line crosses the Y-axis), and
b = the slope of the line.

These relationships are illustrated in Figure 7-1. The value of the *slope*, b, is defined as units of rise per unit of run, so that in this figure, where the line increases two Y units per unit of X, the slope = 2/1 = 2. The *Y-intercept*, a, equals 2. As is apparent from this figure, a knowledge of the coordinates of any two points will determine the precise position of this or any straight line (as will, likewise, the slope and Y-intercept). It was noted earlier that a linear regression could be exemplified by the fitting of a straight line to the points on a scatter diagram. But, it was also noted that the line must be the one that fits the data best. Typically, the points describing any observations are so dispersed because of sampling variation as to make it impossible to connect all of them by a straight line, so that the line of best fit is, not surprisingly, the *running average of all points*. In the case of regression analysis a point on this line also provides the estimate of the mean of the population of Y_i's, $\mu_{Y \cdot X}$, called for by a particular X_i. Fortunately, the linear regression analysis is so constructed as to make it impossible to fit the wrong line to the data unless, of course, a mistake is made in calculation or one attempts to fit a straight line to nonlinear data.

The slope of the regression line is calculated from the formula

$$b = \frac{\sum_i x_i y_i}{\sum_i x_i^2}$$

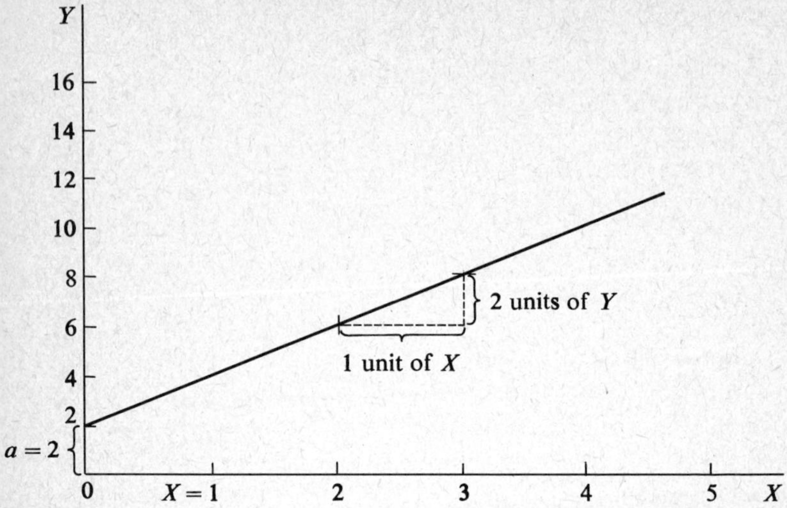

Figure 7-1 An illustration of the basic components of a linear regression.

where x and y are deviations, $X_i - \overline{X}$ and $Y_i - \overline{Y}$, about the means of all observed X's and Y's, respectively. The easiest *computational formula* for this quantity is

$$b = \frac{\sum_i X_i Y_i - \frac{\sum_i X_i \sum_i Y_i}{n}}{\sum_i X_i^2 - \frac{(\sum_i X_i)}{n}}$$

where n = total number of pairs of observations. A positive value for the slope indicates that X and Y increase together; a negative slope indicates that as X increases Y decreases, so that the slope of the line in Figure 7-1 can only be positive.

The Y-intercept for the regression line is computed as

$$a = \overline{Y} - b\overline{X}$$

Logically, there are several ways of rewriting the basic regression equation. Substituting the formula for a in the basic equation, however, provides us with one of the more useful alternative forms. Thus, from

$$\hat{Y}_i = a + bX_i$$

one obtains

$$\hat{Y}_i = \bar{Y} - b\bar{X} + bX_i$$

and

$$\hat{Y}_i - \bar{Y} = -b\bar{X} + bX_i = b(X_i - \bar{X})$$

Thus

$$\hat{y}_i = bx_i$$

The introduction of the new symbolism, $\hat{Y}$ and $\hat{y}$, calls attention to a basic concept of regression. Since these latter formulas describe the regression line—the running average of the observed points—and the Y-coordinate of a point on this line is an estimate of $\mu_{Y \cdot X}$ for a particular value of X, it is likely that many if not the majority of the observed Y_i's will not lie on the regression line. Consequently, the *regression-estimated mean value* for the population of Y_i's corresponding to an X_i, $\hat{Y}_i$ pair must be distinguished from an observed value Y_i. If all $\mu_{Y \cdot X}$'s were known, then the *true* slope of the line, β, would also be known, as would the *true* Y-intercept, α. The regression equation would then become

$$\mu_{Y \cdot X} = \alpha + \beta X_i$$

We are now in a position to begin a regression analysis on data such as those shown in Table 7-1. Several things are immediately obvious from these data: For one, the number of replications for a given dose of X-ray is quite variable, and for another, the increments of dose are not uniform, with five X_i's for doses up to and including 1,000 roentgens but only three observations from 2,000 to 3,000. While the design of this experiment may be less than inspiring, the computations for the regression can be performed without regard to numbers of replicates or spacing of the X_i's. Finally, it is readily apparent that the percentage of lethals does increase with dose, so there is a strong indication that the variables are indeed related. However, one thing that is not immediately obvious from a casual examination of the data is whether the X and Y variables are related in a linear fashion and, therefore, amenable to a linear regression analysis. There are at least three ways to examine this question. One, the data can be plotted on graph paper to confirm (or refute) visually the basic linearity of the relationship. Two, preliminary calculations can be made based on the

Table 7-1 Percent *D. melanogaster* sperm possessing *X*-chromosome recessive lethal mutations after irradiation with X-ray

	X = dose (roentgens)	Y = percent lethals
	500	3.4
	500	3.8
	850	5.8
	1000	5.7
	1000	5.2
	1100	5.6
	1500	7.1
	1500	7.2
	2000	8.0
	2300	9.7
	2950	11.6
	3500	13.4
	3500	13.7
	3500	13.6
	4100	15.7
	4100	15.7
Sum	33,900	145.2
n	16	16
Mean	2,118.8	9.1

knowledge that linearity implies additivity so that the difference in percent lethals between 1,000 and 2,000 roentgens (ca. 2.5%) should be in about the same general order of magnitude as that between 2,000 and 3,000 (ca. 3.6%) and 3,000 to 4,000 (ca. 4.0%). Neither of these ways is foolproof, and if casual inspection were all that dependable there would be little reason to spend the time computing a regression line. For these data the second method provides rather variable results, so that not a great deal of confidence can be placed in it. However, the scatter diagram shown in Figure 7-2 does strongly suggest a linear relationship between radiation dose and frequency of lethal mutations. If you are still in doubt the third, and definitive, test is to compute the linear regression line and the *sum of squares of the deviations of the observations* from it. Then, your second choice of a line is computed; it *must* be curvilinear since you know that the linear regression equation will provide you with the straight line of best fit. The line of best fit

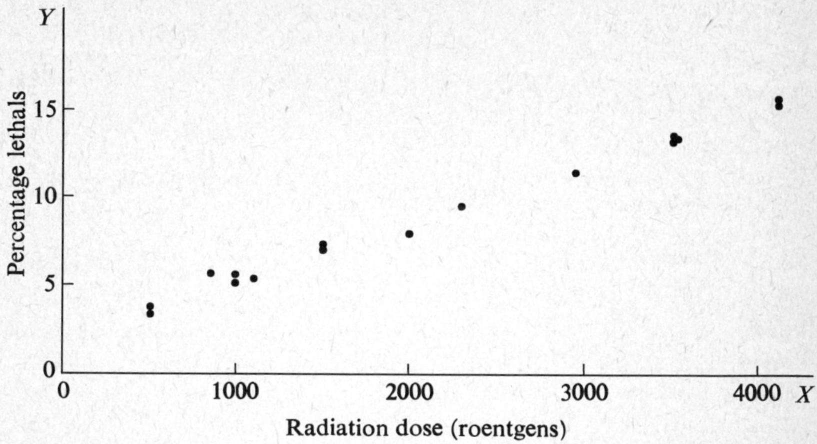

Figure 7-2 Scatter diagram for the data of Table 7-1.

(whether it be linear or curvilinear) will be the one in which this sum of squares is the smallest. Fortunately for the biologist, most biological relationships are linear.

For the data of Table 7-1

$$b = \frac{\sum x_i y_i}{\sum x_i^2} = \frac{83{,}117.5}{25{,}469{,}375} = 0.003$$

and

$$a = \bar{Y} - b\bar{X} = 9.1 - 0.003(2118.8) = 2.7$$

Now $\hat{Y}$, the estimate of $\mu_{Y \cdot X}$ corresponding to any chosen X_i, can be estimated from the formula

$$\hat{Y}_i = 2.744 + 0.003\ X_i$$

Since one measures slope as units rise per unit run (here "rise" is percent lethals and "run" is roentgen units) the value of b must denote "0.003% lethal sperm induced per roentgen unit."

The Y-intercept ($a = 2.7$) is the point where the regression line crosses the Y-axis—that is, where $X = 0$. This means that at 0 roentgens of X-ray there are still about 2.7% lethals observed. Mathematically speaking this is satisfactory, but biologically this is a different matter since zero roentgens of X-ray can hardly have been responsible for the production of any lethal mutations. What is needed here is an examination of the method of data collection in this experiment. It may be that the value $a \neq 0$ simply reflects the background (i.e., "spontaneous") mutation rate to lethality in this organism if the background rate was not known or not subtracted from each observation prior to the analysis. On the other hand, if the spontaneous rate had already been subtracted from each observation then one can only assume that there is a tremendous amount of presently unexplained variability in the observations, as 2.7% cannot be accepted as anything but the crudest estimate of $a = 0$. It happens that the spontaneous mutation rate had not been subtracted from the original observations; in fact, an experimental design such as this provides an estimate of this rate simply by accepting the value of the Y-intercept ($X = 0$) as the spontaneous rate of mutation.

Regression through the origin

In some situations the design of the experiment calls for a regression line through the origin: point (0, 0). In the data of Table 7-1, had the spontaneous rate been known and subtracted from the original observations, it would have been logical to regress the line through the origin since we know beforehand that zero roentgens can induce 0% lethals. The point (0, 0) must lie on the regression line but it is unique in that it does not come from any observation and, therefore, contains no sampling variation. A regression of this type must be treated differently from the previous type although the basic computations are similar. In this case

$$b = \frac{\sum_i X_i Y_i}{\sum_i X_i^2}$$

and, since $a = 0$, a point on the regression line can be plotted from

$$\hat{Y}_i = bX_i$$

Properties of regression

One requirement of regression is that the X_i's be measured without error since if one records a Y value supposedly corresponding to, say, X_1 but in reality belonging to X_2, the mean of the wrong population of Y's will be estimated by the regression. The Y's obviously cannot be measured without (sampling) error although it is assumed that the error population is distributed as described earlier, with mean equal to zero. In most regression analyses it is assumed that all Y populations possess a common variance, although some experiments are designed otherwise.

The regression line itself is characterized in the following way: The point $(\overline{X}, \overline{Y})$ must lie on the regression line. Since the line is a running mean of all observed points and the sum of deviations about an arithmetic mean always equals zero, the sum of deviations about the regression line must equal zero. However, the sum of squares of the deviations of the observed Y_i's from the regression line is the minimum; that is, there is no other line of better fit.

There are two sources of variation in regression. One is attributable to the sampling of different populations of Y's with differing means, $\mu_{Y \cdot X}$'s; the other to sampling variation within the sample (i.e., among the Y_i's drawn from a given population). The former source of variation is denoted as that *due to regression of Y on X* since as the chosen X_i's vary in a regression, so will the $\mu_{Y \cdot X}$'s of the corresponding populations of Y's, provided of course that X and Y are related so that a change in X is accompanied by a change in Y. The latter source is called sampling error, which can be expected to accompany the various Y_i's drawn from a population.

One can perhaps visualize the partitioning of the sources of variation in an anova if the following relationships are noted while consulting Figure 7-3. Any Y_i consists of a sample mean, $\hat{Y}$, and a random deviation $(Y_i - \hat{Y})$, which we will call d. $\hat{Y}$, the sample mean, consists of the grand mean, $\overline{Y}$, and the deviation due to varying the X_i's, $\hat{y}$, such that

$$\hat{y} = \hat{Y} - \overline{Y} = bx$$

By definition,

$$Y_i = \hat{Y} + (Y_i - \hat{Y})$$

Substituting,

$$= \overline{Y} + \hat{y} + (Y_i - \hat{Y})$$

and transposing

$$Y_i - \overline{Y} = \hat{y} + (Y_i - \hat{Y})$$

produces

$$= bx + (Y_i - \hat{Y}) = bx + d_i$$

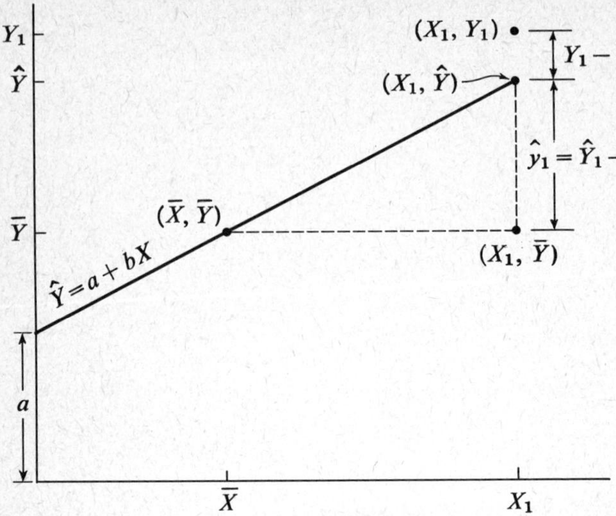

Figure 7-3 Relationships in linear regression.

As a side note, this equation can be rewritten in terms of parameters as

$$Y_i = \mu + \beta x_i + \varepsilon_i$$

where μ is estimated by $\bar{Y}$, β by b, and ε_i by d_i.

Squaring both sides of the sample quantity and summing, one is left with the following *sum of squares for variation in Y*

$$\sum_i (Y_i - \bar{Y})^2 = b^2 \sum_i (X_i - \bar{X})^2 + \sum_i d_i^2$$

Variability due to all sources in regression is measured by

$$\text{Total } SS = \sum_i (Y_i - \bar{Y})^2 = \sum_i Y_i^2 - \frac{(\sum_i Y_i)^2}{n} = \sum_i y_i^2$$

with degrees of freedom equal to $n - 1$.

Variability due to the use of various values of X is measured by

$$\text{Regression } SS = b^2 \sum_i (X_i - \bar{X})^2 = b \sum_i x_i y_i = \frac{(\sum_i x_i y_i)^2}{\sum_i x_i^2}$$

with 1 degree of freedom.

All remaining variability not accounted for by the regression sum of squares is computed as

$$\text{Error } SS = \sum_i d_i^2 = \sum y_i^2 - \frac{(\sum_i x_i y_i)^2}{\sum_i x_i^2}$$

with $n - 2$ degrees of freedom.

For the data of Table 7-1, the components of variability in Y are computed in Table 7-2.

Table 7-2 Regression analysis for the data of Table 7-1

Source of variation	*df*	*SS*	*MS*	*F*
Regression	1	271.25	271.25	2260.42**
Error	14	1.68	0.12	
Total	15	272.93		

The F value is intended to test the significance of the trend in the data—that is, whether a change in X results in a *real* change in Y. If $b = 0$ then there can be no real variability attributable to regression on X and the regression sum of squares will reflect only random deviations about a horizontal regression line. The data of Table 7-1 provide an estimate of β from the statistic $b = 0.003$. As can be seen in the anova summarized in Table 7-2, this slope does indeed appear to differ from $b = 0$ so that $H_0 : b = 0$ is rejected.

The error mean square provides an unbiased estimate of the variability about the regression line and is denoted $s_{Y \cdot X}^2$. Likewise, as the observations in a sample, b as an estimator of β is also subject to sampling error. The *variability in b* can be estimated from

$$s_b^2 = \frac{s_{Y \cdot X}^2}{\sum x^2} = \frac{0.12}{25{,}469{,}375} = 0.00+$$

One conceivably may ask about the degree of confidence that can be placed in any $\hat{Y}_i$ as an estimator of $\mu_{Y \cdot X}$. The confidence limits about a $\hat{Y}_i$ (for example, 95%) may be computed from the formula

$$95\% \text{ confidence } (\hat{Y}_i) = \bar{Y} + b(X_i - \bar{X}) \pm t_{.05}\sqrt{s_{Y \cdot X}^2}\sqrt{\frac{1}{n} + \frac{(X_i - \bar{X})}{\sum_i (X_i - \bar{X})^2}}$$

$$= \bar{Y} + bx_i \pm t_{.05}\sqrt{s_{Y \cdot X}^2}\sqrt{\frac{1}{n} + \frac{x_i}{\sum x^2}}$$

with $n - 2$ degrees of freedom for the tabulated value of Student's t.

Earlier in the discussion of regression we observed that any straight line can be determined by a knowledge of the slope and the Y-intercept. Although any number of lines can have identical values of a, they will differ if $b_1 \neq b_2 \neq \cdots \neq b_n$; likewise, identical slopes do not represent identical lines if $a_1 \neq a_2 \neq \cdots \neq a_n$. However, regardless of the fact that the regression lines in a comparison may differ, a knowledge of the differences or similarities in the slopes can often provide useful information regarding the relationships between the sets of variables examined in the regressions. For example, assume that one regresses percent surviving cells on concentration of bactericide for each of several strains of bacteria. If one knows nothing more than that the b's in the several regressions differ significantly, one may possibly conclude that the modes of action of the chemical on the various cell strains differ; if the b's are equal one may possibly conclude that the basic mode of action is the same throughout all strains. The test used for comparing b's is a t test where

$$t = \frac{b_1 - b_2}{\sqrt{s_p^2[(1/\sum x_1^2) + (1/\sum x_2^2)]}}$$

where s_p^2 denotes the pooled variance of the two regressions, which are in turn distinguished by the subscripts 1 and 2, and

$$s_p^2 = \frac{(\sum y_1^2 - [(\sum x_1 y_1)^2/\sum x_1^2] + \sum y_2^2 - [(\sum x_2 y_2)^2/\sum x_2^2])}{n_1 + n_2 - 4}$$

with $n_1 + n_2 - 4$ degrees of freedom.

This t test is interpreted in precisely the same fashion as the t test on means discussed in Chapter 4; a one- or two-sided hypothesis regarding the difference between the two b's is formulated and either accepted or rejected.

Simple linear correlation

While linear correlation and regression have strong similarities in computation, they are really intended to provide answers to rather different questions. Regression, we saw, enables one to quantify a relationship so that one may, for example, predict the *amount* of increase or decrease of one variable per unit change in the other. *Linear correlation*, on the other hand, provides no means of quantifying a relationship between variables and does not require that one variable be considered dependent upon the other. Instead, it measures the *intensity* of a relationship, so that if any two variables are said to "possess a 90% correlation" it means that as one increases the other will increase (the *amount not specified*) 90% of the time in the long run. This relationship can be expressed in this example as $r =$ *correlation coefficient* $= 0.90$. A negative correlation, $-r$, denotes that as one variable increases the other decreases. Since r denotes a percentage (the sign of r indicates only the direction of the relationship)

$$-1.00 \le r \le 1.00$$

As a general rule there is no need to perform a regression to the exclusion of a correlation analysis, or vice versa. Usually either or both types of analyses can be performed provided the *values for both variables are drawn from individuals who are sampled at random.* If the individuals in a sample are not randomly chosen as regards one variable (for example, dose of radiation as described in Table 7-1, age, time after treatment, locale, etc. are generally not random choices) only the regression analysis can be performed since a correlation analysis generally provides a misleadingly high estimate of the intensity of the relationship between the variables.

It was indicated that the computations for regression and correlation analyses are similar. This should be evident from the formulas

$$r = \frac{\sum xy}{\sqrt{\sum x^2 \sum y^2}} = \sqrt{(b_{Y \cdot X})(b_{X \cdot Y})}$$

where $b_{Y \cdot X}$ and $b_{X \cdot Y}$ denote the slopes of the regressions computed with Y regressed on X and vice versa, respectively. It follows that as

$$b_{Y \cdot X} = \frac{\sum xy}{\sum x^2}$$

that

$$b_{X \cdot Y} = \frac{\sum xy}{\sum y^2}$$

Since no knowledge as to which variable is dependent and which independent is required in a correlation analysis, the labeling of the variables

X and Y is completely arbitrary. When one is applying the formula using the square root of the products of the slopes, the sign of r is the same as that of the two b's since they, in turn, must always possess the same sign.

As was indicated in the case of regression, detecting and measuring linear correlation simply by visual inspection of a scatter diagram is a chancy procedure, too. The wrong choices of scale can obscure a real correlation or synthesize one where none should exist, since the correlation is dependent upon the regression slopes and as the scales change so do the slopes.

Figure 7-4 shows two scatter diagrams. Part (a) shows data for which there is a very low correlation between the variables. Note that the two regression lines, $b_{Y \cdot X}$ and $b_{X \cdot Y}$, tend to lie near the X and Y axes, respectively. In part (b), however, the relationship between the variables is great and the two regression lines lie quite near one another, about midway between the axes.

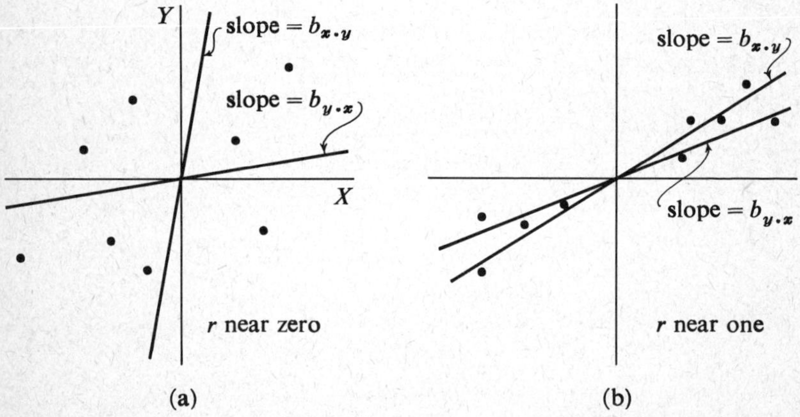

Figure 7-4 Scatter diagrams showing (a) a low degree of linear correlation and (b) a high degree of linear correlation.

Table 7-3 presents data suitable for a correlation analysis. The individuals from which each *pair* of measurements, X_i and Y_i, were obtained were randomly chosen from the population consisting of mature rodents. The conclusion that one may draw from this analysis is

Table 7-3 Oxygen consumption (cc) for mature rodents of different weights (gm)

	Weight (X)	Oxygen consumption (Y)
	198.7	112.6
	154.6	102.1
	231.8	149.5
	165.6	72.8
	193.2	120.3
	207.0	161.5
	212.5	143.5
	115.9	77.8
	132.5	64.3
	160.1	108.2
	190.4	85.1
	267.7	162.2
	204.2	149.5
	124.2	78.1
Mean	182.7	113.4
$\sum x^2$	23,999.4	$\sum y^2$ 15,782.9

$$\sum xy = 16{,}641.5$$

$$\sqrt{\sum x^2 \sum y^2} = 19{,}462.9$$

$$b_{Y \cdot X} = \frac{16{,}641.5}{23{,}999.4} = 0.69$$

$$b_{X \cdot Y} = \frac{16{,}641.5}{15{,}782.9} = 1.05$$

$$r = \frac{16{,}641.5}{19{,}462.9} = 0.85 = \sqrt{0.69(1.05)}$$

that body weight and oxygen consumption are closely related; more specifically, about 85% of the time, as weight increases oxygen consumption can also be expected to increase and, of course, vice versa. Unlike regression, the computed correlation coefficient, r, indicates absolutely nothing as to how much of an increase in one variable can be expected with an increase in the other.

Assuming a random model, one may place equal weight in both regression coefficients. As is seen in the table, a value for $b_{Y \cdot X}$ equal to

0.69 indicates that for each gram of body weight oxygen consumption increases about 0.69 cc. On the other hand, $b_{X \cdot Y}$ denotes that for each cc of oxygen consumed, weight will be expected to increase about 1.05 gm.

When the correlation between two variables is perfect (i.e., $r = \pm 1.00$) the error sum of squares must equal zero. That is, all variability must be attributable to regression. However, the relationship between r and the sum of squares attributable to regression is not linear. The relationship is described by the *coefficient of determination*, r^2, which is quantified as

$$r^2 = \frac{(\sum xy)^2}{\sum x^2 \sum y^2}$$

which in turn equals the *proportion* of the total sums of squares attributable to regression. That is,

$$r^2 = \frac{\text{Regression } SS}{\text{Total } SS}$$

Thus, for the data of Table 7-3 where $r = 0.85$, the amount of variability in the dependent variable due to regression is

$$r^2 = (0.85)^2 = 0.72$$

Thus, about 28% (1.00 – 0.72) of the total variability observed is unexplained.

The sample correlation coefficient, r, is an unbiased estimate of ρ (the Greek letter "rho"), the true correlation coefficient provided that $\rho = 0$; though biased at values of $\rho \neq 0$, r is still the best overall estimate of this parameter. In order to test the hypothesis that $\rho = 0$ (that is, that the observations have been drawn from a bivariate population with no correlation between the variables) one can compute

$$t = \frac{r}{(1 - r^2)/(n - 2)}$$

with $n - 2$ degrees of freedom. For the data of Table 7-3

$$t = \frac{0.85}{(1 - 0.72)/12} = 36.5$$

For this value of t at 12 degrees of freedom, $P << .05$, so that one may conclude that, using r as the estimator, $\rho \neq 0$.

An analysis of variance may be computed for the correlation, which may then be used to provide another method of testing the hypothesis: $\rho = 0$. The F value as computed in Table 7-4 and shown in Table 7-5 also indicates that this parameter is significantly different from zero.

Table 7-4 Anova for testing the significance of the linear correlation coefficient

Source of variation	*df*	*SS*	*MS*	*F*
Regression	1	$\frac{(\sum xy)^2}{\sum x^2}$	SS/df	MS_R/MS_E
Error	$n-2$	$\sum y^2 - \frac{(\sum xy)^2}{\sum x^2}$	SS/df	
Total	$n-1$	$\sum y^2$		

Note also the following equalities
Regression $MS = b^2 \sum x^2 = r^2 \sum y^2$

Table 7-5 Anova for the data of Table 7-3

Source of variation	*df*	*SS*	*MS*	*F*
Regression	1	11,539.4	11,539.4	32.7**
Error	12	4,244.6	353.7	
Total	13	15,783.9		

Though there is every indication that ρ does not equal zero, at this point one has no idea how near to ρ the sample statistic really is. A confidence interval around r (including ρ) would undoubtedly be a very useful bit of information. First off, r must be converted to a probability distribution; it can be converted to approximate normality as a z value. The fit to normality progressively decreases as sample size, n, drops below 100, although generally not enough to cause real problems. Thus,

$$z = 0.5 \ln \frac{1+r}{1-r} = 1.2$$

where ln is the "natural logarithm" ($\log_e$) which, if desired, can be converted to the more generally used $\log_{10}$ as

$$\ln X = \log_e X = 2.3025 \log_{10} X$$

Thus, one can also compute

$$z = 1.1513 \log \frac{1 + r}{1 - r}$$

The widths of the confidence limits around r and including ρ can be obtained from the following relationships.

$$90\% CL = z \pm 1.64\sqrt{1/(n - 3)} = 1.2 \pm 0.49 = .71, .69$$

$$95\% CL = z \pm 1.96\sqrt{1/(n - 3)} = 1.2 \pm 0.59 = .69, .79$$

$$99\% CL = z \pm 2.58\sqrt{1/(n - 3)} = 1.2 \pm 0.77 = .43, .97$$

Since these confidence limits are expressed in terms of z, they must be converted back to r by means of the formula

$$r = \frac{e^{2z} - 1}{e^{2z} + 1} = \frac{2.718_{\dots}^{2z} - 1}{2.718_{\dots}^{2z} + 1}$$

In the current example

$$90\% CL = \frac{2.718_{\dots}^{1.42} - 1}{2.718_{\dots}^{1.42} + 1} \quad \text{to} \quad \frac{2.718_{\dots}^{3.38} - 1}{2.718_{\dots}^{3.38} + 1} = 0.61, 0.94$$

In similar fashion

$$95\% CL = 0.60, 0.95$$
$$99\% CL = 0.41, 0.97$$

and e raised to a power, n, can be obtained from any handbook of statistical or mathematical tables.

Often one wishes to compare two correlation coefficients and determine whether they are the same or differ significantly from one another. Let the null hypothesis be $H_0 : \rho_1 = \rho_2$. While the r's must once again be converted to z values, for this test there is no need to reconstitute the z's back to r values. The test used is, not surprisingly, a t test of the form

$$t = \frac{z_1 - z_2}{\sqrt{1/(n_1 - 3) + 1/(n_2 - 3)}}$$

where

$$z_i = 0.5 \ln \frac{1 + r_i}{1 - r_i} = 1.1513 \log \frac{1 + r_i}{1 - r_i}$$

and the computed value of t is compared to the tabular value of Student's t for a two-sided test with infinite degrees of freedom.

Partial linear correlation

As was discussed earlier at some length, a statistical analysis involving the effects of one or more factors on some metric trait may prove misleading if additional factors that do influence the observations are uncontrolled or are not randomly assigned to each of the observations; that is, if the ε's are not randomly sampled. Fortunately, for fairly large samples the effects of the uncontrolled factors tend to cancel each other out, for the most part. Eventually, uncontrolled variability is assigned to the general category of "sampling error." Such is the case, too, in the correlation analysis. However, there are instances where a correlation between two variables may be grossly misleading because of the effect of the other variables that perhaps were recognized as being capable of effect but were uncontrolled and do not cancel one another out in the observations. Thus, they may cause an over or under estimate of the true correlation between the two variables being examined.

Consider the case where, for example, one wishes to correlate the antibody titer in a milliliter of a mouse's blood—after inoculation with a specific amount of an antigenic substance—to the body weight. Since there is no a priori reason to doubt the validity of this approach, it may provide meaningful information of a relationship between the two variables. On the other hand, age may also be found to be related to antibody titer since younger animals generally are not as able to protect themselves against such antigenic substances as are older ones. Consequently, one might imagine a situation where within any one age group antibody titer may be found positively correlated with body weight but negatively correlated if age is not held constant and if the relationship between age and titer is stronger than that between body weight and titer.

If the problem is recognized beforehand and the number of experimental units suitably abundant, groups of observations may be planned in which at any one time all but two of the variables are held constant.

Sometimes this is more readily said than done so that several variables may, of necessity, have been permitted to vary in the course of the experiment. Provided that the value of each variable under consideration (controlled and uncontrolled) in every experimental unit has been recorded, the observations may be partitioned into the effects of *pairs* of variables.

Using the example of the mouse noted above, simple correlation coefficients are computed in the manner described earlier. We can denote antibody titer as variable 1, body weight as 2, and age as 3. The relationships by correlation can be expressed as

$$\begin{aligned} \text{titer vs body weight} &= r_{12} \\ \text{titer vs age} &= r_{13} \\ \text{age vs body weight} &= r_{23} \end{aligned}$$

Here, r_{12} may be influenced by the varying ages (variable 3), r_{13} by body weight (variable 2), etc. At this point *partial correlation coefficients* may be computed each of which, in effect, provides a measure of the relationship between two variables, the remaining variable(s) being held constant.

The expression

$$r_{12 \cdot 3} = \frac{r_{12} - r_{13} r_{23}}{\sqrt{(1 - r_{13}^2)(1 - r_{23}^2)}}$$

can be read as "the partial correlation coefficient for variables 1 and 2, the effect of variable 3 being held constant"; in essence, the effect of variable 3 is canceled out. The formula for

$$r_{13 \cdot 2} = \frac{r_{13} - r_{12} r_{23}}{\sqrt{(1 - r_{12}^2)(1 - r_{23}^2)}}$$

is obtained from that for $r_{12 \cdot 3}$ simply be transposing 2 and 3 in that formula; $r_{23 \cdot 1}$ is obtained by transposing 1 and 3 in the first formula.

With four or more variables the number of possible combinations logically increases, as does the number of required calculations. For example, for four variables one must initially calculate r_{12}, r_{13}, r_{14}, etc. to obtain $r_{12 \cdot 3}$, $r_{12 \cdot 4}$, $r_{13 \cdot 2}$, etc., which are necessary to obtain, for example,

$$r_{12 \cdot 34} = \frac{r_{12 \cdot 4} - r_{13 \cdot 4} r_{23 \cdot 4}}{\sqrt{(1 - r_{13 \cdot 4}^2)(1 - r_{23 \cdot 4}^2)}}$$

and so on.

The procedure for testing the significance of the partial correlation coefficient is the same as that used for the simple type except that in the former case, degrees of freedom are equal to $n - k - 2$ where $n =$ number of observations in a particular set of data and k is the number of variables held constant in computing the partial coefficient. Thus, for $r_{12 \cdot 34}$, $k = 2$.

Partial linear regression

This type of analysis bears about the same relationship to partial correlation as does simple linear regression to simple correlation. That is, in a multivariate correlation analysis we relate variables 1 and 2, 1 and 3, and so on, in such a manner as to effectively hold all but two variables constant at any one time. *Partial linear regression* measures the unique contribution to an observation, Y_i, made by each of two or more independent variables: $X_1, X_2, \ldots, X_n$. Since this is a linear relationship that we are discussing, it follows that the contributions made to Y_i by the various X's are additive. Taken another way, the population of Y's from which Y_i is drawn is specified by the *sum* of effects of all independent variables on which Y is regressed, simultaneously. This concept can be seen in the *multiple linear regression equation*

$$\hat{Y}_i = a + b_1 X_{1i} + b_2 X_{2i} + \cdots + b_k X_{ki}$$

where $b_1, b_2, \ldots, b_k$ are *partial regression coefficients*. They denote the slope of the contribution to $\hat{Y}_i$ made by the X_1, X_2, etc. classes of variables and may be interpreted as though all X's other than the one described by a particular b have been held constant. This system, based on the principle of a composite of k simple linear regressions, is naturally more complicated and is more difficult to visualize and depict graphically than the regression with but one independent variable. The individual contributions of the X's to the Y's could, taken individually, be depicted by a straight line of Y regressed on X. However, in order to relate each of several X's to a single Y_i, one must depict Y_i as a point on a *plane* whose position in space is ultimately determined by the contribution of each bX. Working with the simplest case of multiple regression—that involving two independent variables, X_1 and X_2—one can construct a three-dimensional figure like that shown in Figure 7-5.

The lowermost plane, $X_1 Y$, considered alone relates the contribution to Y made by independent variable X_1; the uppermost plane,

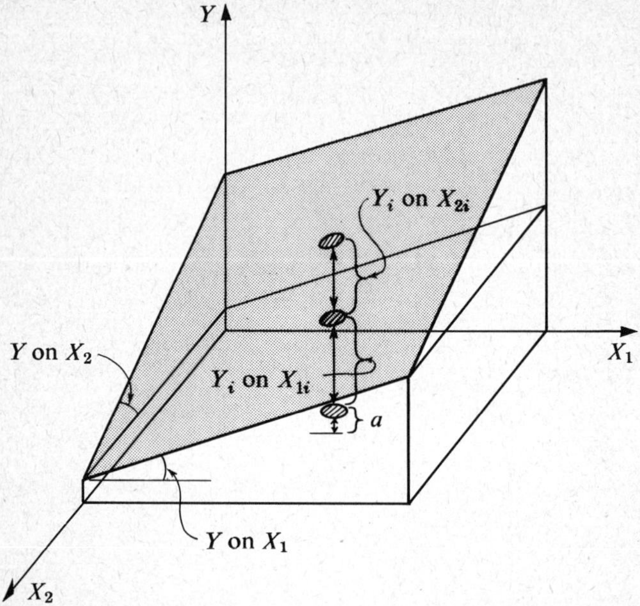

Figure 7-5 The relationships in multiple linear regression.

$X_2 Y$, considered alone relates that contribution made by X_2 while the uppermost surface—that of the X_1, X_2 plane—represents the total effect of both independent variables. Though this figure would be difficult to construct for more than two independent variables, the basic principles are the same regardless of number.

For the case of linear regression with but one independent variable, X, we saw

$$\hat{Y}_i = a + bX_i$$

and since

$$\hat{Y}_i - \overline{Y} = \hat{y}_i$$

the equation could be written as

$$\hat{y}_i = bx_i$$

This last equation can be extended to include the case of k independent variables, so that

$$\hat{y}_i = b_1 x_{1i} + b_2 x_{2i} + \cdots + b_k x_{ki}$$

represents the contribution to an observation *due entirely to regression of Y on k independent variables*.

Earlier, it was also seen that the random deviation of an observed Y_i from the regression was defined as being equal to $Y_i - \hat{Y}_i$, which by substitution becomes $y_i - \hat{y}_i$ and, extended to k independent variables, becomes

$$y_i - \hat{y}_i = y_i - b_1 x_{1i} - b_2 x_{2i} - \cdots - b_k x_{ki}$$

The sum of squares for the deviations from regression, $\sum(Y_i - \hat{Y}_i)^2$, here must also be minimal for the *plane of best fit*. The sum of deviations from regression, $\sum(Y_i - \hat{Y}_i)$, here must also equal zero so that one obtains

$$\sum(Y_i - \hat{Y}_i) = y_i - b_1 x_{1i} - b_2 x_{2i} - \cdots - b_k x_{ki} = 0$$

Transposing,

$$y_i = b_1 x_{1i} + b_2 x_{2i} + \cdots + b_k x_{2k}$$

By multiplying this last equation by x_1, or $x_2, \ldots, x_k$, and summing over all observations, one ends up with

$$\sum_i x_{1i} y_i = b_1 \sum_i x_{1i}^2 + b_2 \sum_i x_{1i} x_{2i} + \cdots + b_k \sum_i x_{1i} x_{ki}$$

and

$$\sum_i x_{ki} y_i = b_1 \sum_i x_{1i} x_{ki} + b_2 \sum_i x_{2i} x_{ki} + \cdots + b_k \sum_i x_{ki}^2$$

which are called *normal equations* and which can be solved simultaneously for the values of $b_1, \ldots, b_k$.

Taking a hypothetical numerical example for a regression with two independent variables X_1 and X_2, let

$$\sum_i x_{1i} y_i = 4 \qquad \sum_i x_{2i}^2 = 1 \qquad \sum_i x_{1i}^2 = 8$$

$$\sum_i x_{2i} y_i = 8 \qquad \sum_i x_{1i} x_{2i} = 7$$

When these are inserted into the two appropriate normal equations

$$4 = b_1 8 + b_2 7$$

and

$$2 = b_1 7 + b_2 1$$

Solving simultaneously one finds

$$b_1 = 0.24$$

and

$$b_2 = 0.30$$

The Y-intercept of a multiple regression analysis equals

$$a = \bar{Y} - b_1\bar{X}_1 - b_2\bar{X}_2 - b_k\bar{X}_k$$

If it is given that $\bar{Y} = 3$, $\bar{X}_1 = 2$, and $\bar{X}_2 = 1$, then

$$a = 3 - 0.48 - 0.30 = 2.2$$

The mean of any population of Y's related to this hypothetical regression may now be estimated from the equation

$$\hat{Y}_i = 2.2 + 0.24X_{1i} + 0.30X_{2i}$$

The determination of b's using the solution afforded by the normal equations is extendable to regressions with any number k of independent variables although the procedure becomes rather tedious as the number of independent variables increases. While it is often suggested in statistical texts that only a computer be used in regressions involving three or more independent variables, generally this only serves to further discourage the biologist. However, there are alternative methods of solution that are less tedious in application than the normal equations and require nothing more sophisticated than a desk calculator. The one to be described here is based on the Gauss-Doolittle method as modified by Steel and Torrie† and is as good as any and better than most. It is based upon the use of *matrices* (ordered *rectangular* arrays of values in which the arrangement is significant) and *vectors* (ordered arrays of values in which the order is significant) but a greater understanding of the rationale of this test is beyond the scope of this book and is not necessary for the successful application of the test.

---

† R. G. D. Steel and J. H. Torrie, *Principles and Procedures of Statistics* (New York: McGraw-Hill, 1960).

The data of Table 7-6 describe observations relating one dependent variable (respiratory rate) to three independent variables (nitrogen, carbon dioxide, and oxygen tension). As a preliminary to the regression analysis certain quantities (e.g., $\sum_i x_{1i}$, $\sum_i x_{2i}$, $\sum_i x_{1i} y_i$, etc.) are calculated. From these values, as shown summarized in Table 7-7, are then generated four matrices and four vectors as shown in Table 7-8. This table forms the basis of the Gauss-Doolittle method, which provides a

Table 7-6 The effect of varying tensions of nitrogen, carbon dioxide, and oxygen on respiratory rate

Nitrogen (X_1)	Carbon dioxide (X_2)	Oxygen (X_3)	Respiratory rate (Y)
5	5	5	36.6
5	5	10	51.7
5	5	15	64.6
5	10	5	25.9
5	10	10	50.6
5	10	15	52.8
5	15	5	22.8
5	15	10	31.9
5	15	15	45.9
10	5	5	32.3
10	5	10	42.1
10	5	15	62.2
10	10	5	20.4
10	10	10	41.0
10	10	15	51.4
10	15	5	13.3
10	15	10	24.9
10	15	15	44.4
15	5	5	31.3
15	5	10	45.7
15	5	15	56.3
15	10	5	18.2
15	10	10	31.9
15	10	15	46.1
15	15	5	9.6
15	15	10	25.3
15	15	15	37.9

Table 7-7 Preliminary calculations accompanying the multiple regression analysis of the data presented in Table 7-6

	X_1	X_2	X_3	Y
n	27	27	27	27
Sum	270	270	270	1017.1
Mean	10	10	10	37.7
$\sum X^2$	3,150	3,150	3,150	43,921.4

$\sum_i X_{1i} X_{2i} = 2{,}700$ $\sum_i X_{1i} X_{3i} = 2{,}700$ $\sum_i X_{2i} X_{3i} = 2{,}700$

$\sum_i X_{1i} Y_i = 9{,}768.5$ $\sum_i X_{2i} Y_i = 9{,}337.6$ $\sum_i X_{3i} Y_i = 11{,}427.0$

$$\sum_i x_{1i}^2 = \sum_i X_{1i}^2 - \frac{(\sum_i X_{1i})^2}{n} = 3{,}150 - \frac{72{,}900}{27} = 450$$
$$\sum_i x_{2i}^2 = 450$$
$$\sum_i x_{3i}^2 = 450$$

$$\sum_i x_{1i} x_{2i} = \sum_i X_{1i} X_{2i} - \frac{\sum_i X_{1i} \sum_i X_{2i}}{n} = 2700 - \frac{72{,}900}{27} = 0$$
$$\sum_i x_{1i} x_{3i} = 0$$
$$\sum_i x_{2i} x_{3i} = 0$$

$$\sum_i x_{1i} y_i = \sum_i X_{1i} Y_i - \frac{\sum_i X_{1i} \sum_i Y_i}{n} = 9{,}768.5 - \frac{274{,}617.0}{27} = -402.5$$
$$\sum_i x_{2i} y_i = -833.4$$
$$\sum_i x_{3i} y_i = 1{,}256.0$$

stepwise solution to the *three* normal equations required to solve this regression problem.

Matrix X is obtained from the sums of products for the x's:

$$a_{11} = \sum_i x_{1i}^2 \qquad a_{12} = \sum_i x_{1i} x_{2i} \quad \cdots \quad a_{33} = \sum_i x_{3i}^2.$$

Some of the values in this matrix are redundant (e.g., $\sum_i x_{1i} x_{2i} = \sum_i x_{2i} x_{1i}$) and so are deleted. Vector Y is derived from sums of products for x_i and y:

$$a_1 y = \sum_i x_{1i} y_i \quad \cdots \quad a_3 y = \sum_i x_{3i} y_i.$$

Table 7-8 Matrices and vectors necessary for solving a multiple regression problem as exemplified by the data of Table 7-6

	Matrix X				*Vector Y*
	x_1	x_2	x_3		y
x_1	$a_{11} = 450$	$a_{12} = 0$	$a_{13} = 0$	x_1	$a_1 y = -402.5$
x_2		$a_{22} = 450$	$a_{23} = 0$	x_2	$a_2 y = -833.4$
x_3			$a_{33} = 450$	x_3	$a_3 y = 1256.0$
	Matrix X − 1				*Vector Y − 1*
	$A_{11} = 450$	$A_{12} = 0$	$A_{13} = 0$		$A_1 y = -402.5$
	1	$B_{12} = 0$	$B_{13} = 0$		$B_1 y = -0.89$
	Matrix X − 2				*Vector Y − 2*
		$A_{22} = 450$	$A_{23} = 0$		$A_2 y = -833.4$
		1	$B_{23} = 0$		$B_2 y = -1.8$
	Matrix X − 3				*Vector Y − 3*
			$A_{33} = 450$		$A_3 y = 1256.0$
			1		$B_3 y = 2.8$

Matrix $X - 1$ can be generated as follows: Copy the values in the first row in Matrix X (a_{11}, a_{12}, etc.) and record them as A_{11}, A_{12}, etc. in this new matrix. Then, each of the A's is divided by a_{11} to produce the B values (e.g., $B_{13} = A_{13}/a_{11}$). Vector $Y - 1$ is developed in similar fashion: $a_1 y$ becomes $A_1 y$; then, $A_1 y/a_{11} = B_1 y$.

Entries in Matrix $X - 2$ are developed from a basic equality. As an example, $A_{2\textcircled{2}} = a_{2\textcircled{2}} - A_{12} B_{1\textcircled{2}}$; $A_{23} = a_{2\textcircled{3}} - A_{12} B_{1\textcircled{3}}$. Each circled subscript must agree throughout the equation. To obtain the B's for this matrix, each A value is divided by A_{22} ($B_{22} = A_{22}/A_{22}$, etc.). The $Y - 2$ vector is similarly developed: $A_{2\textcircled{y}} = a_{2\textcircled{y}} - A_{12} B_{1\textcircled{y}}$ which value is also divided by A_{22} to yield $B_2 y$.

Matrix $X - 3$ is obtained from the following type of equation: $A_{3\textcircled{3}} = a_{3\textcircled{3}} - (A_{13} B_{1\textcircled{3}} + A_{23} B_{2\textcircled{3}})$. The B entries are calculated by dividing the A entries by A_{33} (e.g., $B_{33} = A_{33}/A_{33}$). Vector $Y - 3$ is computed in analogous fashion as $A_{3\textcircled{y}} = a_{3\textcircled{y}} - (A_{13} B_{1\textcircled{y}} + A_{23} B_{2\textcircled{y}})$. $B_3 y$ is computed by dividing $A_3 y$ by A_{33}.

The slopes of the contributions made to Y by each independent variable X_1, X_2, X_3 can now be fairly readily computed from the following relationships:

$$b_3 = B_3 y = 2.79$$
$$b_2 = B_2 y - B_{23} b_3 = -1.85$$
$$b_1 = B_1 y - B_{12} B_2 - B_{13} B_3 = -0.89$$

Any $\hat{Y}_i$ can now be obtained from the equation

$$\hat{Y}_i = 37.22 - 0.89X_{1i} - 1.85X_{2i} + 2.79X_{3i}$$

where

$$a = \bar{Y} - b_1\bar{X}_1 - b_2\bar{X}_2 - b_3\bar{X}_3 = 37.22$$

As noted earlier, this system can be extended to k independent variables by enlarging the matrices and vectors. Each additional variable, for example, would require equivalent enlargement of matrix X ($\sum_i x_{1i}x_{4i}$, etc.) and Vector Y ($a_4\,y$) and would also require an additional matrix, $X - 4$, and vector, $Y - 4$.

The significance of the portion of the total sums of squares attributable to regression of Y on X_1, X_2, and X_3 can be tested by an anova where

$$\text{Total } SS = \sum_i (Y_i - \bar{Y})^2 = \sum_i y^2$$

$$\text{Regression } SS = b_1 \sum_i x_{1i} y_i + \cdots + b_k \sum_i x_{ki} y_i$$

$$\text{Error } SS = \text{Total } SS - \text{Regression } SS$$

The results of the anova performed on the data of Table 7-6 are tabulated in Table 7-9. The variability attributable to regression (Regression MS) is seen to be highly significant ($P < .005$). Error MS is equivalent to $s^2_{y \cdot 123}$: variability of the estimate, Y, regressed on X_1, X_2, and X_3.

There is one additional matrix, C, also called the "backward solution" of the Gauss-Doolittle method that should be noted since it provides a method by which tests of significance of the slopes may

Table 7-9 Anova performed on the data of Table 7-6

Source of variability	*df*	*SS*	*MS*	*F*
Regression	$k = 3$	5,409.1	1,803.0	209.7**
Error	$n - k - 1 = 23$	197.8	8.6	
Total	$n - 1 = 26$	5,606.9		

be calculated. Without describing its origin, let it suffice to list the following equalities:

$$C_{44} = 1/A_{44} \qquad C_{34} = -B_{34}C_{44} \qquad C_{24} = -B_{23}C_{34} - B_{24}C_{44}$$
$$C_{14} = -B_{12}C_{24} - B_{13}C_{34} - B_{14}C_{44}$$

$$C_{33} = 1/A_{33} \qquad C_{23} = -B_{23}C_{33} - B_{24}C_{43}$$
$$C_{13} = -B_{12}C_{23} - B_{13}C_{33} - B_{14}C_{43}$$

$$C_{11} = 1/A_{11}$$

and

$$C_{23} = C_{32}\text{, etc.}$$

Using the quantities computed from these equations one can estimate the *variance of a slope*, $s_{b_k}^2$, such that

$$s_{b_1}^2 = C_{11}s_{y\cdot 123}^2 = C_{11}MS_E$$
$$s_{b_k}^2 = C_{kk}\,s_{y\cdot 123}^2 = C_{kk}\,MS_E$$

and the *standard deviation* which is given by

$$s_{b_k} = \sqrt{s_{b_k}^2}.$$

To test the null hypothesis, $H_0 : b_k = 0$, one uses a t test where

$$t = \frac{b_k}{s_{b_k}}$$

with degrees of freedom equal to $n - k - 1$.

To test the *equality of two slopes*—that is, $H_0 : b_j = b_k$—one must first calculate the standard deviation of the difference between the two b's of interest. For example,

$$s_{(b_1 - b_2)} = \sqrt{MS_E(C_{11} + C_{22} - 2C_{12})}$$

A t test is also applied here, where

$$t = \frac{b_1 - b_2}{s_{(b_1 - b_2)}}$$

with degrees of freedom equal to $n - k - 1$.

PROBLEMS

1. Nine medical students were randomly chosen from a larger group in order to determine whether performance on the midterm could be used to predict

performance on the final examination. When the two sets of scores are plotted on equal interval paper the resulting scatter diagram suggests a linear relationship.

Student	*Midterm score*	*Final score*
1	27	50
2	36	61
3	36	53
4	37	44
5	25	52
6	18	47
7	17	37
8	32	58
9	25	52

(a) Plot the two sets of variates on equal interval graph paper and draw a line of best fit estimated by eye. (b) Compute the linear regression equation for these data. (c) Predict the final grade for an individual receiving a grade of 26 on the midterm. (d) Predict the midterm score for a student who received a 54 on the final examination. (e) Draw the regression line on the graph made in (a) and compare it to the line fitted by eye.

2. Since the data of Problem 1 represent randomly drawn pairs of variates, it is proper to compute a linear correlation coefficient, r. Using these data compute this value and interpret the result.

3. The following data represent the dry weight (mgm), X, and height (cm), Y, of randomly chosen representatives of a single species of plant.

X	Y
4.2	129.7
5.8	180.3
9.7	270.4
4.5	130.2
7.5	269.2
6.0	121.4
5.6	170.1
7.2	187.6
8.4	249.2
7.0	200.5
7.7	239.2
4.8	107.2
6.9	141.9
7.4	249.1
4.1	100.8

(a) Make a scatter diagram of these values and estimate by eye the line of best fit. (b) Calculate the linear regression coefficients, $b_{y\,.\,x}$ and $b_{x\,.\,y}$. (c) Compute the values of $a_{y\,.\,x}$ and $a_{x\,.\,y}$ and complete the linear regression equations

$$\hat{Y} = a_{y\,.\,x} + b_{y\,.\,x} X \quad \text{and} \quad \hat{X} = a_{x\,.\,y} + b_{x\,.\,y} Y$$

(d) Fit the two computed regression lines to the scatter diagram. (e) Calculate r. (f) Test the hypothesis that $\rho = 0$.

4. In the following data, four Y measurements were obtained for each X.

$X =$	3.0	2.5	2.0	1.5	1.0	0.5
	6.6	5.9	5.0	4.9	4.2	2.8
	6.4	5.8	5.6	4.5	3.8	3.6
$Y =$	7.2	6.2	5.1	4.1	4.6	3.0
	6.8	5.4	5.2	4.6	4.1	3.2

(a) Calculate $b_{y\,.\,x}$. (b) Test the hypothesis that $\beta = 0$. (c) Can a test of linear correlation be made? Why?

5. Given that $r_{12} = 0.795$, $r_{13} = -0.046$, $r_{23} = 0.355$, (a) compute the partial correlation coefficients, $r_{12\,.\,3}$, $r_{13\,.\,2}$, and $r_{23\,.\,1}$. (b) For each of these six coefficients, test the hypothesis that $\rho_i = 0$ given that $N = 28$ for these data. (c) By use of the z transformation compute a 0.95 confidence limit about r_{12} and $r_{12\,.\,3}$.

6. Given that

$$\sum x_1^2 = 9.846 \qquad \sum x_1 y = -5.823$$
$$\sum x_1 x_2 = 2.141 \qquad \sum x_2 y = -4.216$$
$$\sum x_2^2 = 10.621 \qquad \sum y^2 = 6.689$$

(a) compute r_{12}, r_{y1}, r_{y2}. (b) Using the values obtained in part (a) compute $r_{y1\,.\,2}$ and $r_{y2\,.\,1}$.

7. Given the following,

$$\sum x_{1i} y_i = 6 \qquad \sum x_{2i}^2 = 4 \qquad \sum x_{1i}^2 = 12$$
$$\sum x_{2i} y_i = 24 \qquad \sum x_{1i} x_{2i} = 9$$

solve for the values of partial regression coefficients, b_1 and b_2.

8. In an experiment the following data were obtained:

X_1	X_2	Y
1	0	2
2	0	3
2	0	2
3	1	5
4	0	5
5	1	7
6	0	5
8	1	6
9	1	7
10	1	8

Compute b_1, b_2, and the Y-intercept, a.

ANSWERS TO PROBLEMS

1. (b) $\hat{Y} = 35.3 + 0.54X$
2. $r = 0.981$
3. (b) $b_{y.x} = 32.38$ $\quad b_{x.y} = 0.024$
   (e) $r = 0.877$
   (f) $SS_{TOT} = 51{,}095.16$ $\quad SS_{REG} = 39{,}270.45$ $\quad SS_E = 11{,}824.72$
   $t = 6.58$
4. (a) $b = 1.35$
   (b) $F = 1.06$
   (c) No, as the X's and Y's were not randomly drawn.
5. (a) $r_{12 \cdot 3} = 0.867$ $\quad r_{13 \cdot 2} = -0.584$ $\quad r_{23 \cdot 1} = 0.648$
   (b) r_{12}: $P < 0.001$ $\quad r_{13}$: $P > 0.05$ $\quad r_{23}$: $P > 0.05$
   $r_{12 \cdot 3}$: $P < 0.001$ $\quad r_{13 \cdot 2}$: $P < 0.01$ $\quad r_{23 \cdot 1}$: $P < 0.001$
6. (a) $r_{12} = 0.209$ $\quad r_{y1} = -0.718$ $\quad r_{y2} = -0.499$
   (b) $r_{y1 \cdot 2} = -0.724$ $\quad r_{y2 \cdot 1} = -0.513$
7. $6 = b_1 12 + b_2 9 \twoheadrightarrow 18 = b_1 36 + b_2 27$
   $24 = b_1 9 + b_2 4 \twoheadrightarrow 96 = b_1 36 + b_2 16$
   $b_2 = 7.09$ and $b_1 = -4.82$
8. $b_1 = 0.42$ $\quad b_2 = 1.52$ $\quad a = 2.14$

8

ANALYSIS OF COVARIANCE

The analysis of *covariance* is an often unappreciated but nonetheless extremely useful procedure that combines the techniques of linear regression and analysis of variance in *correcting for heterogeneity among experimental units.* More specifically, this technique is applied to data in order to correct or adjust the effects attributable to the principal variable under consideration, Y, because of variability in some other factor, X, which may have been recognized during the experiment but for one reason or another was not or could not be controlled to the extent that it could be neatly incorporated into a statistical analysis like, say, a two-way factorial.

Assume, for example, that you wish to determine the relative effectiveness of two antimalarial agents in individuals suffering from malaria, drawn from among patients in a military hospital. While each of the patients is assigned to one of the two treatment categories, one may find that there is considerable heterogeneity in the duration of time a person has been afflicted: It may vary from weeks or months to years. The effect of this heterogeneity in duration of affliction may be suspected of affecting the individual's response to the treatment but, due to the mechanics of the data collection, it would most likely be impossible to find adequate numbers of individuals who are all identical in this regard.

Consequently, in a case such as this the measure of the *effectiveness* of the malarial treatment, *variable* Y, can be adjusted due to the varying values of *duration of affliction*: the *concomitant* or *uncontrolled variable*, X.

Consider two populations: one is representative of the various values of X and the other, the various values of Y. The mean of the population of *differences*, $X_i - Y_i$, is given as

$$\mu_d = \frac{\sum_i d_i}{n}$$

while the variance of the population of differences is found as

$$\sigma_d^2 = \sum_i \frac{(d_i - \mu_d)^2}{n - 1}$$

or else, as

$$= \sigma_X^2 + \sigma_Y^2 - 2\omega$$

The Greek letter "omega," ω, symbolizes the parameter covariance and is computed as

$$\omega = \sum_i \frac{(X_i - \mu_x)(Y_i - \mu_y)}{n - 1}$$

and

$$w = \sum_i \frac{(X_i - \bar{X})(Y_i - \bar{Y})}{n - 1}$$

as the statistic covariance.

As is evident from the latter two equations, covariance is analogous to variance except that the latter is described in terms of only one variable, X or Y, while the former includes both. In fact, if these last two equations were written only in terms of X, or Y, they would equal, precisely, the equations for variance.

If ω (or its estimator, w) is zero the variables under scrutiny are judged independent. Thus, one need have no fear of the concomitant variable affecting the results. If ω is positive or negative the two variables are related—that is, correlated since covariance can be used as a measure of correlation.

There is no one accepted method for correcting the Y_i's because of the varying values of X since this correction depends entirely on the relationship between the X's and the Y's. If the relationship is known, the adjustment may be made directly without resorting to an analysis of covariance; in most experiments, however, there is no obvious or known relationship between the two, and thus it must be determined from the observations of the experiment itself.

The basic assumption underlying the analysis of covariance is that the variability observed in the concomitant variable is not due to the treatments themselves; for example, duration of affliction is not due to antimalarial drug treatment. A starting point for the analysis is the initial though generally unproven assumption of a *linear* relationship between X's and Y's. In addition, if there is any relationship between the variables, Y must be assumed to be dependent on X since if it were the reverse, the values of X could not affect the observed Y's. Thus, if one assumes

$$Y_i = a + bX_i$$

then one must also assume that

$$Y_{ij} = \mu + \beta_i + \tau_j + \varepsilon_{ij}$$

which says that an observation, Y_{ij}, consists of a population mean (μ), a component due to variability in X (β_i), one due to the effect of the treatment (τ_j), and a random (sampling) error term (ε_{ij}).

The data of Table 8-1 describe the results of blood samples taken from among sixteen individuals afflicted with one type of malarial parasite. Each individual is scored on the basis of mean number of parasites found in twenty erythrocytes after receiving one of two treatments, Y_{ij}; likewise, X_{ij} is the estimated number of months elapsed since this individual contracted the disease.

For these data

$$\text{Covariance} = w_1 = \frac{-11.2}{7} = -1.6$$

for treatment 1, and for treatment 2

$$w_2 = \frac{-80.3}{7} = -11.5$$

Here the quantities for both treatments are negative although covariance may be positive or negative. Since the observed covariances are not equal to zero, it appears possible that the two variables—mean number of parasites and months elapsed since initial infection—may be negatively correlated. However, quantitatively there is little else that one can do with these values of w, so that the quantity, covariance, is rarely calculated.

The *analysis of covariance* is far more useful and involves the partitioning of the *total sum of products* described as

$$\text{Total } SP = \sum_{i,j} X_{ij}\, Y_{ij} - \frac{\sum_i X_{i.} \sum_i Y_{i.}}{\sum_i r_i}$$

Table 8-1[a]

	Treatment 1					Treatment 2				
	X_{1j}	Y_{1j}	X_{1j}^2	Y_{1j}^2	$X_{1j}Y_{1j}$	X_{2j}	Y_{2j}	X_{2j}^2	Y_{2j}^2	$X_{2j}Y_{2j}$
	5	7.8	25	60.8	39.0	6	19.3	36	372.5	115.8
	7	18.3	49	334.9	128.1	9	18.4	81	338.6	165.6
	26	9.2	676	84.6	239.2	13	17.1	169	292.4	222.3
	6	17.1	36	292.4	102.6	13	14.4	169	207.4	187.2
	17	13.2	289	174.2	224.4	19	12.1	361	146.4	229.9
	6	11.2	36	125.4	67.2	8	14.6	64	213.2	116.8
	20	19.5	400	380.2	390.0	7	17.9	49	320.41	125.3
	5	14.5	25	210.2	72.5	18	12.5	324	156.2	225.0
Sum	92	110.8	1536	1663.0	1263.0	93	126.3	1253	2047.1	1387.9
Mean	11.5	13.8				11.6	15.8			
$\sum_{i,j} x_{ij}y_{ij} =$	-11.2					-80.3				

[a] Each Y_{ij} denotes the mean number of malarial parasites in 20 erythrocytes sampled from one patient, while X_{ij} denotes the number of months since initial nfection. Treatments 1 and 2 are two antimalarial treatments.

where X_{ij} and Y_{ij} denote the jth observation of X or Y in the ith treatment, so that for the present example, $i = 1, \ldots, t = 2$ and $j = 1, \ldots, r = 8$ in each statement.

For the data of Table 8-1

$$\text{Total } SP = (1263.0 + 1387.9) - \frac{(92 + 93)(110.8 + 126.3)}{16}$$
$$= -90.6.$$

One component of total variability is that due to treatment. The treatment sum of products can be obtained as

$$\text{Treatment } SP = \sum_i \frac{X_{i.}\, Y_{i.}}{r_i} - \frac{\sum_i X_{i.} \sum_i Y_{i.}}{\sum_i r_i}$$
$$= \frac{92(110.8)}{8} + \frac{93(126.3)}{8} - \frac{(92 + 93)(110.8 + 126.3)}{16}$$
$$= 0.9$$

The error sum of products can be obtained by subtraction, as

$$\text{Error } SP = \text{Total } SP - \text{Treatment } SP$$
$$= -90.6 - 0.9 = -91.5$$

If one makes the not unreasonable assumption that the regression of Y on X is not influenced by treatment—that is, that the concomitant variable affects all individuals in all treatments in precisely the same fashion—an overall slope of the regression line can be estimated from

$$b = \frac{\text{Error } SP}{\text{Error } SS_x} = \text{change in } Y/\text{unit } X$$

where Error SS_x is the error sum of squares for the concomitant variable and can be calculated as

$$\text{Error } SS_x = \sum_{i,j} X_{ij}^2 - \sum_i \frac{(X_{i.})^2}{r_i}$$
$$= 1536 + 1253 - \frac{(92)^2}{8} + \frac{(93)^2}{8}$$
$$= 649.9$$

Thus, for these data

$$b = \frac{-91.5}{649.9} = -0.14$$

which, taken at face value, says that the mean number of parasites decreases *spontaneously* by about 0.14 per month after initial infection.

As was discussed in Chapter 7, a point on the regression line can be described relative to its deviation from that vertical coordinate, $\bar{Y}$; that is,

$$\hat{y}_i = bx_i$$

The observed value of the Y variable, Y_i, in this experiment can now be *adjusted* due to a knowledge of the effect of the concomitant variable so that

$$Y'_{ij} = Y_{ij} - bx_{ij}$$

where Y'_{ij} represents the new value of Y_{ij}, adjusted for the effect of X, duration of infection.

The purpose of this experiment was to analyze the effect of treatment—that is, H_0: $\tau_1 = \tau_2$ or H_0: $\mu_1 = \mu_2$—but in order to do this some additional calculations must be made.

For example, the treatment means, $\bar{Y}_i$'s, must be adjusted as

$$\bar{Y}'_i = \bar{Y}_i - b(\bar{X}_{i.} - \bar{X}_{..})$$

so that

$$\bar{Y}'_1 = 13.8 - (-.14)(11.5 - 11.6) = 13.8$$

and

$$\bar{Y}'_2 = 15.8 - (-.14)(11.6 - 11.6) = 15.8$$

so that for the data of this particular example, the adjusted treatment means are virtually the same as the unadjusted statistics.

However, continuing the analysis of covariance one can additionally compute *adjusted* sums of squares and mean squares for the Y variables as follows:

$$\text{Total } SS'_y = \sum_{i,j} Y_{ij}^2 - \frac{Y_{..}^2}{\sum_i n_i} - \frac{(\text{Total } SP)^2}{\text{Error } SS_x}$$

$$= (1662.9 + 2047.1) - \frac{(110.8 + 126.3)^2}{16} - \frac{(-90.6)^2}{649.9}$$

$$= 183.9$$

$$\text{Error } SS'_y = \sum_{i,j} Y_{ij}^2 - \sum_i \frac{Y_{i.}^2}{n_i} - \frac{(\text{Error } SP)^2}{\text{Error } SS_x}$$

$$= (1662.9 + 2047.1) - \frac{(110.8)^2}{8} + \frac{(126.3)^2}{8} - \frac{(-91.5)^2}{649.9}$$

$$= 168.6$$

with $\sum r - t - 1 = 13$ degrees of freedom.

$$\text{Treatment } SS'_y = \text{Total } SS'_y - \text{Error } SS'_y$$
$$= 183.9 - 168.6 = 15.3$$

with $t - 1 = 1$ degree of freedom.

The test of the hypothesis of equal adjusted treatment means, $H_0\colon \bar{Y}'_1 = \bar{Y}'_2$, is obtained from

$$F = \frac{\text{Treatment } MS'_y}{\text{Error } MS'_y} = \frac{15.3/1}{168.6/14} = 1.3$$

with 1 and 13 degrees of freedom.

The F distribution corresponding to this value indicates that the probability of observing a value of F greater than 1.3 is in excess of 10%. Thus, the null hypothesis of no difference between treatments is accepted.

Table 8-2 summarizes the computations necessary for performing the analysis of covariance.

PROBLEMS

1. Shown below are the average vitamin A content (mgm), Y, and dry weights (mgm), X, of two groups of carrots. Since it was not possible to obtain all plants at the same stage of maturity, the dry weight was used as an estimate of maturity (i.e., covariate). Each group represents a different carrot variety.

Group 1		*Group 2*	
X	Y	X	Y
0	3	2	2
1	5	3	4
3	5	5	4
4	6	6	5
4	8	6	7
6	9	8	8

Table 8-2 Analysis of covariance

Source of variability	SP	SS'_y	df	F
Treatment	$\sum_i \frac{X_{i.} Y_{i.}}{n_i} - \frac{\sum_i X_{i.} \sum_i Y_{i.}}{\sum_i n_i}$	Total SS'_y − Error SS'_y	$t-1$	$\frac{\text{Treatment } MS'_y}{\text{Error } MS'_y}$
Error	Total SP − Treatment SP	$\sum_{i,j} Y_{ij}^2 - \sum_i \frac{Y_{i.}^2}{n_i} - \frac{(\text{Error } SP)^2}{\text{Error } SS_x}$		$\sum r - t - 1$
Total	$\sum_{i,j} X_{ij} Y_{ij} - \frac{\sum_i X_{i.} \sum_i Y_{i.}}{\sum_i n_i}$	$\sum_{i,j} Y_{ij}^2 - \frac{Y_{..}^2}{\sum_i n_i} - \frac{(\text{Total } SP)^2}{\text{Error } SS_x}$		

Error $SS_x = \sum_{i,j} X_{ij}^2 - \sum_i \frac{(X_{i.})^2}{n_i}$

(a) Compute the analysis of covariance. (b) Do you accept the hypothesis $H_0: \mu_1 = \mu_2$?

2. A group of 20 freshman medical students were classified according to duration of premedical schooling and as to whether they had graduated from an in-state or out-of-state premedical college. Their grades after completion of their first anatomy examination were recorded on a four-point scale ($A = 4$, $B = 3$, etc.).

	3-year	*4-year*
In-state	2 1 3 0 1	4 3 4 1 2
Out-of-state	2 0 3 0 1	3 2 2 1 2

Convert the student's period of premedical schooling to a binary by adding one to each 3-year and zero to each 4-year score in the study. (a) Perform an analysis of covariance letting the school of origin be the independent variable, X; the grade the dependent variable, Y, and duration of schooling the variable to be controlled. (b) Test the hypothesis $H_0: \mu_{in} = \mu_{out}$.

ANSWERS TO PROBLEMS

1. (a) $\bar{X}_1 = 3$ $\quad Y_1 = 6$ $\quad \bar{Y}_1' = 6.92$
   $\bar{X}_2 = 5$ $\quad Y_2 = 5$ $\quad \bar{Y}_2' = 4.08$
   (b) $F = 22.6$ (1, 9 *df*) Thus, H_0 rejected.
2. (b) $F = 1.45$ (1, 17 *df*) $P > 0.05$

9

CURVILINEAR REGRESSION

Any biological data displaying a relationship between the dependent and independent variables can apparently be fitted to a linear regression line, even if the variables are *not* linearly related. Committing this type of error is costly, however, since describing a curvilinear relationship with a straight line does not produce the line of best fit (i.e., where the sum of the squares of the deviations, $\sum(\hat{Y}_i - Y_i)^2$, is the minimum); but, more important, some portion of the relationship between the variables (that is, variability in Y truly attributable to regression on X) is lost and is simply relegated to the class of "unexplained" (error) variability. As it happens, the task of fitting a curved line to the observations, or even determining that a curved line is needed, is not always simple or precise. This is not meant to discourage the reader but simply to alert him to some of the tribulations of these sorts of analyses. For example, most nonlinear relationships between variables can be described by several equations, each of which possesses approximately the same goodness of fit to the data. However, since the equations may vary substantially in form, they may require different biological interpretations, so that it is generally wise to have some sort of biological hypothesis in mind when finally choosing the equation that will be used to describe the present observations as well as to predict future ones. Unfortunately the biological interpretation of the terms in the chosen regression equation sometimes poses problems. For example, if the equation has a logarithmic or an X^2 term, to cite two possibilities, it can be difficult to read a structural or biological meaning into these terms. Why, indeed, is it that the amount of change in Y per unit in X is not constant but varies directly with the values of X chosen? Oftentimes one does not have a ready answer and must be content viewing such quantities as presently unexplainable but necessary to the goodness of fit of the line.

At the risk of appearing overly simplistic, it is worth noting that there are two basic approaches to line fitting for nonlinear data. One is called *direct* and involves fitting the data to the line of best fit, usually expressed as some polynomial ("many termed") function. The other is by the *transformation* of the curvilinearly related data to linearity and subsequently plotting the points as a straight line. While logarithmic transformation is the most common type of transformation, others such as square root and probit transformations are possible alternatives.

Polynomials

These relationships can be

linear	$\hat{Y}_i = a + bX_i$
quadratic	$\hat{Y}_i = a + bX_i + cX_i^2$
cubic	$\hat{Y}_i = a + bX_i + cX_i^2 + dX_i^3$

and so on. Examples of this type of function are shown in Figure 9-1, parts (a), (b), (c). These functions are handled in essentially the same way as was outlined in Chapter 7 except that here X_i, corresponding to the pair of variables, X_i, Y_i, can be raised to successively higher powers. For powers of X other than unity (i.e., the linear equation), the line described by the equation is curvilinear. Polynomial functions have peaks and depressions (maxima and minima) numbering one less than the highest power of X, so that a cubic polynomial has one maximum and one minimum for a total of two. While the equations $\hat{Y}_i = a + bX_i$ and $\hat{Y}_i = a + bX_i + cX_i^2$, for example, have the coefficients a and b in common, in practice $a_i \neq a_j$, $b_i \neq b_j$, etc., so that new values for the coefficients must be computed each time another X^n term is added to the equation.

The basic procedure for fitting curvilinear data to a polynomial of unknown form is first to fit the data to a linear regression and compute the sum of squares attributable to regression on X in the way described in Chapter 7. Then, one adds the quadratic (X^2) term by solving the new equation, $\hat{Y}_i = a + bX_i + cX_i^2$, and computing the sum of squares attributable to regression on X and X^2 [that is, $SS(X, X^2)$]. If it is then determined that $SS(X, X^2)$ is significantly greater than the $SS(X)$ obtained from the linear regression, one knows that he has taken some of the variability initially assigned to

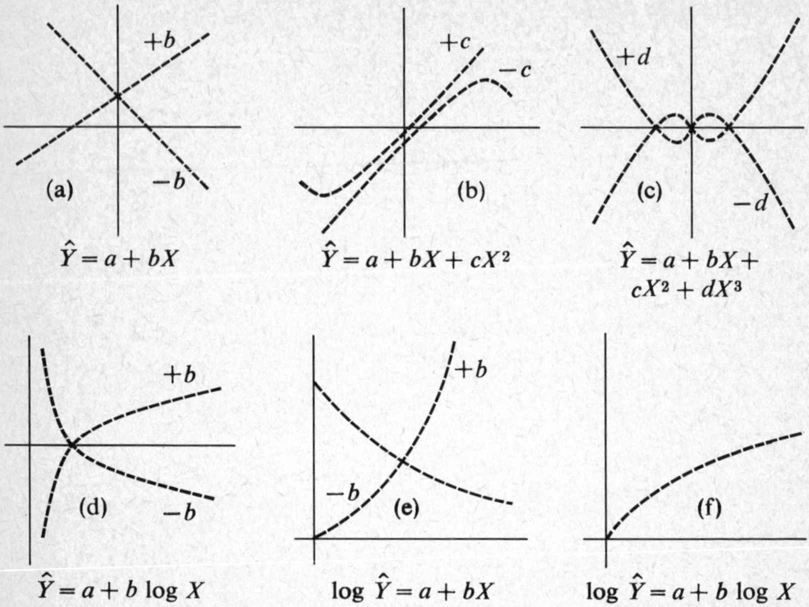

Figure 9-1 Some general types of linear and nonlinear relationships.

error in the linear function and reassigned it correctly to regression, via the quadratic term. This procedure can continue by adding the X^3 term, etc., until a point is reached where no more of the error variability is able to be assigned to regression on X.

Fitting the quadratic polynomial to the data as noted earlier begins with first fitting the data to a linear regression. Though this produces a value for a, b, and the regression sum of squares, the calculation of the quadratic equation requires that a and b, etc., be recomputed taking into consideration the addition of the quadratic coefficient c. For the quadratic,

$$b = \frac{[\sum_i (X_i - \overline{X})(Y_i - \overline{Y})][\sum_i (X_i^2 - \overline{X_i^2})^2] - [\sum_i (X_i^2 - \overline{X_i^2})(Y_i - \overline{Y})][\sum_i (X_i - \overline{X})(X_i^2 - \overline{X_i^2})]}{[\sum_i (X_i - \overline{X})^2][\sum_i (X_i^2 - \overline{X_i^2})^2] - [\sum_i (X_i - \overline{X})(X_i^2 - \overline{X_i^2})]^2}$$

and

$$c = \frac{[\sum_i (X_i - \overline{X})^2][\sum_i (X_i^2 - \overline{X_i^2})(Y_i - \overline{Y})] - [\sum_i (X_i - \overline{X})(X_i^2 - \overline{X_i^2})][\sum_i (X_i - \overline{X})(Y_i - \overline{Y})]}{[\sum_i (X_i - \overline{X})^2][\sum_i (X_i^2 - \overline{X_i^2})^2] - [\sum_i (X_i - \overline{X})(X_i^2 - \overline{X_i^2})]^2}$$

where

$$\sum_i (X_i - \overline{X})(X_i^2 - \overline{X_i^2}) = \sum_i X_i^3 - \left(\sum_i X_i\right)\left(\sum_i X_i^2\right) \Big/ n$$

$$\sum_i (X_i^2 - \overline{X_i^2})^2 = \sum_i X_i^4 - \left(\sum_i X_i^2\right)^2 \Big/ n$$

$$\sum_i (X_i^2 - \overline{X_i^2})(Y_i - \overline{Y}) = \sum_i X_i^2 Y_i - \left(\sum_i X_i^2\right)\left(\sum_i Y_i\right) \Big/ n$$

When these quantities are computed the quadratic equation can be written

$$\hat{Y}_i = \overline{Y} - b\overline{X} + bX_i - c\overline{X_i^2} + cX_i^2$$

inserting the computed values for b and c. As the equation for the Y-intercept, a, is

$$a = \overline{Y} - b\overline{X} - c\overline{X^2}$$

this quantity can be readily determined and also substituted into the equation rewritten in the familiar form

$$\hat{Y}_i = a + bX_i + cX_i^2$$

For the quadratic

$$\text{Total } SS = \sum_i (Y_i - \overline{Y})^2 = \sum_i y_i^2$$

with degrees of freedom equal to $n - 1$.

The quadratic regression sum of squares—that is, the portion of total variability attributable to regression on X and X^2—is

$$\text{Regression } SS = b \sum_i x_i y_i + c \sum_i (X_i{}^2 - \overline{X_i^2}) y_i = SS(X, X^2)$$

with 2 degrees of freedom.

The unexplained quadratic sum of squares is computed as

$$\text{Error } SS = \sum_i y_i^2 - b \sum_i x_i y_i + c \sum_i (X_i^2 - \overline{X_i^2}) y_i$$

with degrees of freedom equal to $n - 3$.

In order to determine whether the fit of the line to the data is improved by the addition of the X^2 term, the *difference* between the linear and quadratic regression sums of squares is computed as

$$SS(X, X^2) - SS(X) = SS(X^2)$$

with one degree of freedom, where $SS(X)$ is the familiar linear regression sum of squares and $SS(X^2)$ denotes the amount of the total sums of squares attributable to the X^2 term.

The test of significance of this reduction in total sums of squares due to the addition of the quadratic term is performed as

$$F = \frac{MS(X^2)}{MS_E}$$

where MS_E is the quadratic error mean square.

As might be expected, the calculation of polynomials of a higher order $(X^3, X^4, \ldots, X^n)$ becomes increasingly complicated although it is by no means a hopeless task without the aid of a computer. We saw earlier how, when Y was regressed on several different X's in multiple regression (recall Chapter 7), one could solve for the coefficients of the X's by means of a method like that of Gauss-Doolittle. In exactly analogous fashion one can use any such multiple regression procedure in a polynomial (curvilinear) regression situation except that X, X^2, X^3, etc., are now used as the X inputs instead of the previous $X_1, X_2, X_3, \ldots, X_k$.

Logarithmic transformation

This type of transformation is the one most commonly applied to biological data since it most often converts curvilinear observations to linearity. Logarithmic functions are generally characterized by a flattening of the curve at one end of the range when plotted on equal interval paper [see Figure 9-1, parts (d), (e), (f)] and become linear when plotted on the appropriate type of logarithmic graph paper. (*Log–log* paper has both X and Y axes graduated on the logarithmic scale while the *semilogarithmic* type has one axis graduated on the log scale and the other on the equal interval scale.)

Most frequently it is the dependent variable, Y, that is transformed by logarithmic equations; this type of transformation is indicated when *percentage* changes in Y vary directly (that is, linearly) with the values of X chosen. Consider, for example, data that appear to follow a logarithmic format when plotted as a scatter diagram on equal interval (regular) graph paper. The first step in determining the correct equation for describing the observations might be to plot the points on semilogarithmic graph paper with the Y coordinates of the observed points being plotted on the logarithmic scale.

Consider the data of Table 9-1 as a case in point. Here the biomass, Y, of cultures of a unicellular organism is recorded as a function of the age, X, of the culture. As seen in Figure 9-2, the relationship between biomass and age is curvilinear when plotted on equal interval graph paper, but linear when plotted on semilogarithmic paper—the Y values being plotted on the logarithmic scale. Since the curvilinear relationship becomes linear after this graphing transformation, one may conclude that the appropriate regression equation for these data is

$$\log \hat{Y}_i = a + bX_i.$$

Table 9-1 Total biomass of cultures of a unicellular organism, as a function of the age of the cultures

X = Age (days)	Y = Biomass (mg)	log of Y
1	72	1.857
2	92	1.964
3	110	2.041
4	137	2.137
5	168	2.225
6	210	2.322
7	263	2.420
8	321	2.506
9	379	2.579
10	471	2.673

$\sum_i X_i = 55$ $\qquad (\sum_i X_i)^2 = 3025$

$\bar{X} = 5.5$ $\qquad \sum_i \log Y_i = 22.724$

$\sum_i X_i^2 = 385$ $\qquad \sum_i X_i (\log Y_i) = 132.430$

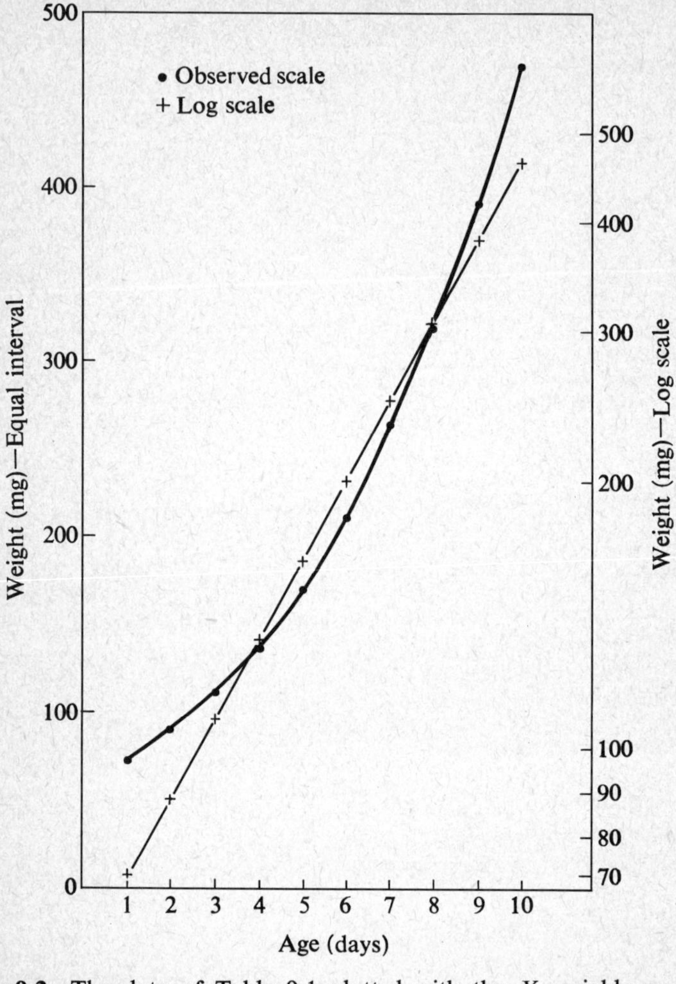

Figure 9-2 The data of Table 9-1 plotted with the Y variable on equal interval and logarithmic scales.

Since this equation converts curvilinearly related data to linearity and since the exponent of X is unity, not surprisingly the methods of solving the regression equation are analogous to those described earlier for the linear regression. For example, for the linear regressions described in Chapter 7 one computed

$$b = \frac{\sum_i X_i Y_i - (\sum_i X_i)(\sum_i Y_i)/n}{\sum_i X_i^2 - (\sum_i X_i)^2/n} = \frac{\sum_i x_i y_i}{\sum_i x_i^2}$$

while for the logarithmic function just described

$$b = \frac{\sum_i X_i(\log Y_i) - (\sum_i X_i)(\sum_i \log Y_i)/n}{\sum_i X_i^2 - (\sum_i X_i)^2/n} = \frac{\sum_i x_i \log y_I}{\sum_i x_i^2}$$

In similar fashion, the Y-intercept, a, for this function is found as

$$a = \frac{\sum_i \log Y_i}{n} - b\bar{X} = \overline{\log Y_i} - b\bar{X}$$

where the values of Y_i are converted to logarithms before the computations are begun.

For the data of Table 9-1

$$b = \frac{132.430 - 55(22.724)/10}{385 - 3025/10} = \frac{7.448}{82.5}$$

$$= .090 \text{ mg on log scale}$$

and

$$a = \frac{22.724}{10} - .090(5.5) = 1.777 \text{ mg on log scale}$$

The slope of the regression in this example is interpreted as "change of 0.090 mg in logarithmic units per day increment in age." The Y-intercept is also measured in logarithmic units on the Y-axis.

If an examination of the data indicates that percent changes in the independent variable produce direct, or linear, changes in Y then a transformation of the sort

$$\hat{Y}_i = a + b \log X_i$$

is indicated. Obviously, one should confirm this conclusion by graphing on both regular and semilogarithmic paper with the X coordinates of the observations being plotted on the logarithmic scale. If the relationship becomes linear when plotted in the fashion just described, then one may proceed with the solution for

$$b = \frac{\sum_i (\log X_i)(Y_i) - (\sum_i \log X_i)(\sum_i Y_i)/n}{\sum_i (\log X_i)^2 - (\sum_i \log X_i)^2/n}$$

and

$$a = \bar{Y} - b\frac{\sum_i \log X_i}{n}$$

Again it should be noted that the conversion of X to logarithms must be done before the calculations are undertaken. The slope in this case may be represented as "change of b units of Y per logarithmic unit increase in X."

Data that appear to follow a relationship like that of the function

$$\log \hat{Y}_i = a + b \log X_i$$

may be transformed to linearity by the relationships of

$$b = \frac{\sum_i (\log X_i)(\log Y_i) - (\sum_i \log X_i)(\sum_i \log Y_i)/n}{\sum_i (\log X_i)^2 - (\sum_i \log X_i)^2/n}$$

and

$$a = \frac{\sum_i \log Y_i}{n} - b\frac{\sum_i \log X_i}{n}$$

where b may be interpreted as "logarithmic units of change in Y per logarithmic unit increase in X."

After completing a transformation regression it is perfectly correct to discuss the relationship between the observations in terms of, say, the logarithm of the value; unfortunately it is generally very hard to visualize changes in one variable as a function of the logarithm of the other. Consequently, it may be advantageous in the interpretation of the data if a slope or Y-intercept computed in logarithmic units, or any other units of transformation for that matter, is also expressed in equal interval units by means of computing, say, the antilog of the values under discussion.

The quadratic curve fitted to logarithms

Sometimes data are observed that, even when plotted on semilogarithmic paper, continue to show a slight curvature—that is to say, a slight concavity either upward or downward as is shown in Figure 9-3. The problem becomes one of determining the equation that best describes the relationship between the X and Y variables. Oftentimes the

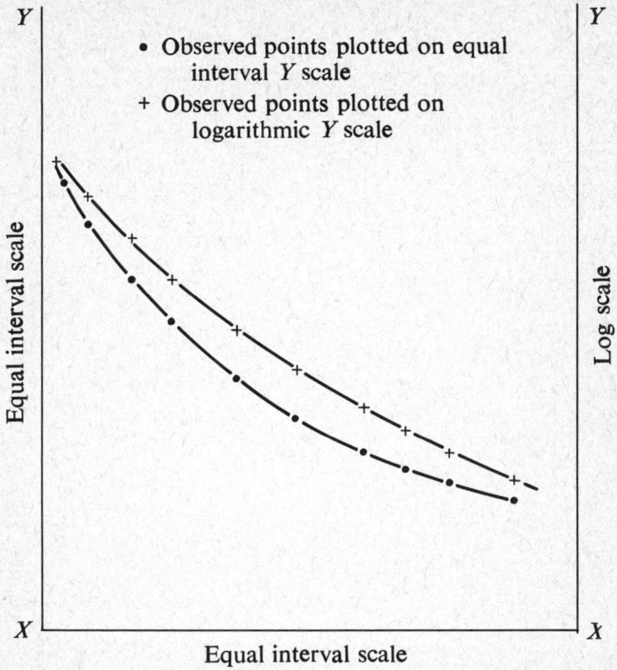

Figure 9-3 A case where the line of best fit only approaches linearity when plotted semilogarithmically.

fitting of a *quadratic polynomial to the logarithms of the Y values* is the solution. The relationship is described as

$$\log \hat{Y}_i = \log a + X_i \log b + X^2 \log c$$

where

$$\log b = \frac{[\sum_i (X_i - \bar{X})(\log Y_i - \log \bar{Y})][\sum_i (X_i^2 - \overline{X_i^2})^2] - [\sum_i (X_i^2 - \overline{X_i^2})(\log Y_i - \log \bar{Y})][\sum_i (X_i - \bar{X})(X_i^2 - \overline{X_i^2})]}{[\sum_i (X_i - \bar{X})^2][\sum_i (X_i^2 - \overline{X_i^2})^2] - [\sum_i (X_i - \bar{X})(X_i^2 - \overline{X_i^2})]^2}$$

and

$$\log c = \frac{[\sum_i (X_i - \bar{X})^2][\sum_i X_i^2 - \overline{X_i^2})(\log Y_i - \log \bar{Y})] - [\sum_i (X_i - \bar{X})(X_i^2 - \overline{X_i^2})][\sum_i (X_i - \bar{X})(\log Y_i - \log \bar{Y})]}{[\sum_i (X_i - \bar{X})^2][\sum_i (X_i^2 - \overline{X_i^2})^2] - [\sum_i (X_i - \bar{X})(X_i^2 - \overline{X_i^2})]^2}$$

The following relationships are useful in computing the coefficients of the quadratic equation:

$$\sum_i (X_i - \bar{X})(X_i^2 - \overline{X_i^2}) = \sum_i X_i^3 - \left(\sum_i X_i\right)\left(\sum_i X_i^2\right)\Big/ n$$

$$\sum_i (X_i^2 - \overline{X_i^2})^2 = \sum_i X_i^4 - \left(\sum_i X_i^2\right)^2 / n$$

$$\sum_i (X_i^2 - \overline{X_i^2})(\log Y_i - \log \bar{Y}) = \sum_i X_i^2 \log Y_i - \left(\sum_i X_i^2\right)\left(\sum_i \log Y_i\right)\Big/ n$$

The log of the Y-intercept a can be computed from

$$\log a = \log \bar{Y} - \bar{X} \log b - X^2 \log c$$

The asymptotic growth curve: a modified exponential function

The equation for a straight line, $\hat{Y}_i = a + bX_i$, describes a constant *amount* of increase or decrease in Y, while the *exponential* equation of the form

$$\hat{Y}_i = ab^{x_i}$$

describes a constant *ratio* of change in Y. If $b > 1$ the trend is upward and the amount of change is a constant percent increase, while if $0 < b < 1$ the trend is downward and the change is a constant percent decrease.

The equation

$$\hat{Y}_i = k + ab^{x_i}$$

is a *modified exponential* equation that describes a relationship in which the absolute amount of growth declines but never reaches—only approaches—a limit called the *asymptote*.

The four basic asymptotic curves are shown in Figure 9-4. For example, part (a) of this figure might represent the case where the absolute number of individuals afflicted with a genetic disease increases as a constant percentage of population size, while part (c) might represent the declining number of deaths due to affliction with a specific disease, as a percent function of, say, time. In the modified exponential equation, k represents the value of the asymptote; a, the vertical distance between Y and the asymptote when $X = 0$, and b is the constant measure of change: it is analogous to the slope of the line.

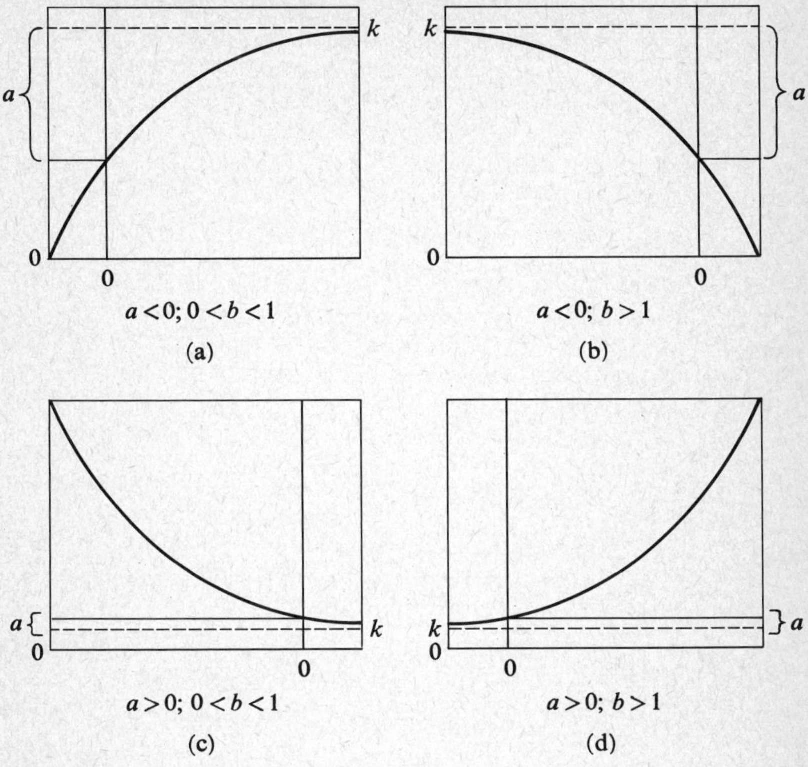

Figure 9-4

One will note that there are no curves denoting values of $b < 0$, since such values are meaningless.

Table 9-2 presents data that represent the numbers of afflicted individuals per 1,000 population with $X = 0$ denoting the base value for these data—in this case, the value of X for 1,000 individuals. Thus, $X = 1$ represents 2,000 individuals, $X = 2$ represents 3,000, etc. These data when plotted are analogous to the curve shown in Figure 9-4(a). It can be seen from an examination of the table that the number of afflicted individuals in the population has a reasonably constant percentage *decrease* in amount of growth; that is, though each successive Y value is larger than its predecessor, the relative trend is one of decline as evidenced by the decreasing magnitudes of changes in Y, ΔY.

Table 9-2 Numbers of afflicted individuals (Y) per 1,000 population (X)

Interval	X	Y	Partial total of Y	ΔY	Percent of preceding increment
I	0	55	—	—	—
	1	73	128	18	—
II	2	86		13	0.72
	3	96	182	10	0.77
III	4	103		7	0.70
	5	109	212	6	0.85

This trend of decline can also be deduced from the last column of the table, percentage of preceding increment, where each value is seen to be only about 75% larger than its predecessor. It is also worthwhile to note that the observations shown in the table are divided into three sections with equal numbers of X's. This is a necessary computational requirement for the solution of the equations describing this function. The formulas to be described here apply equally well to all asymptotic functions.

The rate of change is found as

$$b^n = b^2 = \frac{\sum_{\text{III}} Y - \sum_{\text{II}} Y}{\sum_{\text{II}} Y - \sum_{\text{I}} Y} = \frac{212 - 182}{182 - 128} = \frac{30}{54} = 0.56$$

$$b = \sqrt{b^2} = 0.75$$

where n represents the number of thousands of individuals per each one-third of the data. This value of $b = 0.75$ may be interpreted as the relative rate of decrease in rate, per unit increase in X. Continuing,

$$a = \sum_{\text{II}} Y - \sum_{\text{I}} Y \frac{b - 1}{(b^n - 1)^2} = 54 \frac{-0.25}{0.19} = -69.35$$

and

$$k = \frac{1}{n}\left[\sum_{\text{I}} Y\left(\frac{b^n - 1}{b - 1}\right)a\right] = \frac{1}{2}\left[128\left(\frac{-0.44}{-0.25}\right)(-69.35)\right] = 124.85$$

If it is desired to solve for k without first solving for a and b, the equation

$$k = \frac{1}{n}\left[\frac{(\sum_{\text{I}} Y)(\sum_{\text{III}} Y) - (\sum_{\text{II}} Y)^2}{\sum_{\text{I}} Y + \sum_{\text{III}} Y - 2\sum_{\text{II}} Y}\right]$$

can be used. Thus, the equation describing these data is

$$\hat{Y}_i = 124.85 - 69.35(0.75)^{x_i}$$

so that for

$$\begin{aligned} X &= 0, & \hat{Y} &= 55.5 \\ X &= 1, & \hat{Y} &= 73.0 \\ &\vdots & &\vdots \\ X &= 5, & \hat{Y} &= 108.7 \end{aligned}$$

The graph of the data described in Table 9-2, is shown in Figure 9-5.

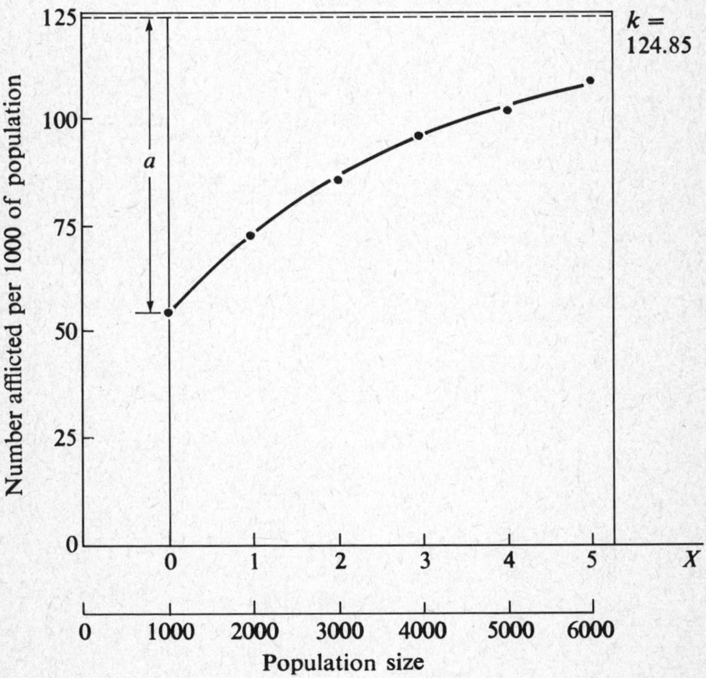

Figure 9-5 A representation of the data of Table 9-2. (See text for full explanation.)

PROBLEMS

1.

X	Y
1	2
2	3
2	2
3	5
4	5
5	7
6	5
8	6
9	7
10	8

(a) Plot these paired variates on a scatter diagram. Note the difficulty in deciding whether a line fitted by eye should be linear or slightly curvilinear. (b) Compute a linear regression equation for these data. (c) In that the curve has at most a single bend, it appears that a quadratic equation would be the type most likely to improve on the fit of the linear type if, indeed, an improvement is possible. Compute a quadratic regression equation. (d) Compute $SS(X)$ and $SS(X, X^2)$ for these data. Has the addition of the X^2 term significantly improved the fit of the line?

2. The following data describe the rate of linear movement (mm/min) of a terrestrial insect as a function of ambient temperature (°C).

X	Y
10	40
11	60
12	60
12	60
13	63
14	64
14	84
15	85
15	84
16	87
16	88
17	88
19	100
20	110
21	112

(a) Plot the data on equal interval paper and by eye determine the line of best fit. How many bends are there in the line? Do the data seem to follow a polynomial or a logarithmic form? (b) Plot the data on semilogarithmic paper. Do the data become linear if either X or Y is on the log scale? (c) State the general form of the regression equation that best describes the data.

3.

X	Y	X	Y
76.2	24.0	106.3	55.7
76.8	22.0	106.8	58.0
77.3	26.5	119.0	61.4
79.2	22.0	119.7	69.0
80.0	25.0	119.8	71.5
87.8	37.5	119.9	73.0
93.2	36.0	123.5	76.5
93.5	39.5	141.0	78.5
94.3	32.0	142.3	74.0
96.8	26.5	145.5	71.5
97.5	55.5	149.0	77.0
99.6	49.8	150.0	85.5
103.2	56.1		

Determine the regression line of best fit and compute the exact regression equation.

4. Consider the following data:

X	Y	log Y	X	Y	log Y
−16	277.2	2.443	1	713.8	2.854
−15	255.4	2.407	2	631.0	2.800
−14	226.4	2.355	3	576.5	2.761
−13	168.0	2.225	4	558.1	2.747
−12	161.8	2.209	5	554.4	2.744
−11	191.6	2.282	6	568.8	2.755
−10	219.1	2.341	7	592.7	2.773
− 9	258.6	2.413	8	605.1	2.782
− 8	291.1	2.464	9	596.8	2.776
− 7	286.4	2.457	10	628.5	2.798
− 6	305.8	2.485	11	641.3	2.807
− 5	318.1	2.502	12	649.9	2.813
− 4	390.3	2.591	13	658.0	2.818
− 3	464.2	2.667	14	697.9	2.844
− 2	411.6	2.614	15	697.6	2.844
− 1	444.9	2.648	16	694.7	2.842
0	477.2	2.679			

Consider first only the pairs of variates: X, Y. (a) Plot the values of X and Y on equal interval paper. Are they linear? (b) Now plot the points on semilogarithmic paper, first with X and then with Y on the log scale. Notice that the data seem to be least curvilinear when Y is plotted logarithmically. What sort of function does the latter pattern suggest? (c) Treat the data now by fitting the logarithmic function to a quadratic one. Compute the exact quadratic regression equation that fits these data.

ANSWERS TO PROBLEMS

1. (b) $\hat{Y} = 2.05 + 0.59X$
   (c) $\hat{Y} = 0.79 + 1.25X + 0.06X^2$
2. (c) $\log \hat{Y} = a + bX$
3. $\hat{Y} = -138.5 + 2.7X - 0.008X^2$
4. $\log \hat{Y} = 2.662 + 0.019X - 0.0004X^2$

10
DETERMINATION OF DISTRIBUTIONS AND RELATIONSHIPS

In our previous discussions of statistical tests as applied to biological data it was usually implied though not always stated that the majority of such are *assumed* to have come from at least approximately normally distributed populations. Indeed, most statistical tests are based on this assumption. However, in previous chapters we have often acknowledged the existence of distributions other than the normal; the chi square, binomial, Student's t, etc., are but a small sample of the many different ones in existence. Not surprisingly, then, there will be times when the usual assumption of normality will be in doubt either as regards the applicability of a particular statistical test that requires this assumption or in making predictions about future observations. Furthermore, even if the data clearly are not normal it is always prudent to know the probable type (i.e., distribution) of population from which the observations were drawn. While the classification of a distribution by inspection is sometimes possible it can, too, be misleading, so that generally one needs to qualify the judgment of distribution type with a specified degree of confidence. This can be accomplished by measuring the goodness of fit of the observations to the expectations associated with a certain type of distribution.

Normal distribution

As was noted in an earlier chapter there are an infinity of normal distributions, each uniquely determined by μ and σ. Thus, when testing data for normality one must ask if the observations fit *the* normal curve appropriate to the observations. For illustration, consider the weight of a particular visceral organ of the human body (see Table 10-1). In this table the weights of the organs sampled have been recorded only to the nearest gram so that the 81-gm class, for example, includes all observations within the interval 80.50–81.49 gm. Noting the symmetry of the observed numbers—approximately symmetrical tails, modal class at about the center of the range, etc.—the data certainly look as though they could have come from a normal distribution, but though this might be true our aim here is to accept or reject this notion not by inspection but by a specific degree of confidence in the decision. The rationale that will be applied here is quite simple and is as follows: The observed estimates of μ and σ, $\bar{X}$ and s from the sample, will provide the statistics necessary for computing a normal curve. Then, the observations will be compared to the values expected in a normal distribution characterized by these values of $\bar{X}$ and s and measure the goodness of fit of the observations to the expected values.

Table 10-1 Observed weights of visceral organs (gm) fitted to a normal distribution

Weight (gm)	Interval	Normal probability	Expected no.	Observed no.
71	70.50–71.49	0.0108	1.32	1
72	71.50–72.49	0.0231	2.82	4
73	72.50–73.49	0.0444	5.42	6
74	73.50–74.49	0.0758	9.25	9
75	74.50–75.49	0.1065	12.99	13
76	75.50–76.49	0.1376	16.79	18
77	76.50–77.49	0.1465	17.87	19
78	77.50–78.49	0.1393	16.99	16
79	78.50–79.49	0.1156	14.10	14
80	79.50–80.49	0.0829	10.11	9
81	80.50–81.49	0.0533	6.50	5
82	81.50–82.49	0.0283	3.45	5
83	82.50–83.49	0.0133	1.62	2
84	83.50–84.49	0.0055	0.67	1

$\bar{X} = 77.11$
$N = 122$
$s^2 = 7.07$
$s = 2.66$

Earlier were noted the advantages of converting particular normal distributions to the common z distribution with $\mu = 0$ and $\sigma = 1$; it is to our advantage to do this here, also. However, as was noted earlier, it is nearly impossible to discuss one discrete point in any continuously variable distribution. Just as the observed values lie in intervals 1 gram wide—simply because they could not be measured more accurately than ± 0.5 gm—the expected values must also be set up in intervals of fixed width. The size of the interval chosen is usually not critical but it should be intuitively obvious that the smaller the intervals the greater their number and the smoother and more precise the shape of the curve. Consequently, we will set up the expected intervals with the minimum permissible width for these data (1 gm) as shown in the second column of this table.

The normal probabilities of column three are computed directly from the standard z distribution, where the probability of any observed value of weight lying in any interval can be computed from

$$P(X_i < X < X_j) = P\left(\frac{X_i - \mu}{\sigma} < z < \frac{X_j - \mu}{\sigma}\right)$$

or, using the data of the present example,

$$\begin{aligned} P(70.50 < X < 71.49) &= P\left(\frac{70.50 - 77.11}{2.66} < z < \frac{71.49 - 77.11}{2.66}\right) \\ &= P(2.48 < z < 2.11) \\ &= P(z < 2.11) - P(z < 2.48) \\ &= 0.0108 \end{aligned}$$

After the normal probability for each interval has been computed, the expected *numbers* of observations in an interval can be determined very simply as

$$\text{Expected number} = \text{normal probability} \times 122$$

For example, for these data, $0.0108 \times 122 = 1.32$ organs are expected to be found in the interval 70.50–71.49 gm drawn from a normal distribution with $\bar{X} = 77.11$ and $s = 2.66$. As can be seen by inspection, at this point the observed and expected values seem to agree fairly well, strongly suggesting that the hypothesis: Both the expected and observed distributions of values are of the same type, is correct. However, the goodness of fit has yet to be quantified. The test most commonly applied to this end is the chi square test. This distribution, although mentioned in an earlier chapter, has yet to be defined. At this point one should note that

$$\chi^2 = \sum \frac{(\text{Observed} - \text{Expected})^2}{\text{Expected}}$$

with observed and expected denoting values of the variable under study. The degrees of freedom for this test are equal to the number of classes or groups (intervals, in the context of the data of Table 10-1) *less* the number of restrictions: the number of constants used to compute the expected values that were obtained from the observations themselves. In the current example the restrictions are three in number, due to the use of n, $\overline{X}$, and s to this end. Whenever the expected number in a class is less than 1, the eventual chi square value is an overestimate of the true value. This discrepancy can be minimized, however, by combining consecutive expected classes so that no such class is less than 1. It is also worth noting here that a chi square test can *never* be computed on data —either observed or expected—that are expressed as a percentage of the total: the values used must always represent actual numbers of observed or expected individuals.

For these data

$$\chi^2 = \frac{(1 - 1.32)^2}{1.32} + \frac{(4 - 2.82)^2}{2.82} + \cdots + \frac{[(2 + 1) - (1.62 + 0.67)]}{1.62 + 0.67}$$

$$= 2.241 \text{ with } 13 - 3 = 10 \text{ degrees of freedom}$$

Since the last class in Table 10-1 (84 gm) has an expected number of only 0.67, the last two classes were combined for the test and, therefore, reduced the degrees of freedom from 14 – 3 to 13 – 3. As can be seen from Table C in the Appendix, the probability of obtaining a larger χ^2 value, by chance alone, is about 99%. Thus the difference between the observed values and those expected in a normal distribution is not judged significant so that it may safely be assumed that the distribution of organ weights in this study is normal.

Binomial distribution

This type of discrete distribution describes systems where there are only two possible outcomes to an observation: Yes or No, On or Off, Present or Absent, etc. Any probability of occurrence in this type of distribution is constant from trial to trial, sample to sample. For example, in the cases of sexes of children in families the probability of birth of a boy or a girl does not change regardless of the number of children already born in that family or their classification as regards sex. Since this distribution was described at some length in an earlier chapter, let it suffice to reiterate that the complete array of outcomes for any binomial distribution of n observations per sample can be obtained from the binomial expansion expression

$$(p+q)^n$$

where p and q represent the probabilities of occurrence of the two outcomes. We also saw that one could compute the probability of finding *exactly* X "favorable" occurrences (i.e., occurrences of interest) of an event in samples of n observations from the expression

$$P_x = \frac{n!}{X!(n-X)!} p^x (1-p)^{n-x}$$

where p and q are defined exactly as they were immediately above.

The goodness of fit of data to a binomial distribution can be determined in a fashion essentially analogous to that of the normal distribution just described. Table 10-2 describes a hypothetical example in which the observed numbers of favorable events in $N = 79$ samples, each consisting of $n = 5$ observations, are scored. Logically, the number of favorable events per sample can range from zero to five. The probability of occurrence of the favorable event, p, in this example is assumed

Table 10-2 Fitting observations to a hypothetical binomial distribution

Favorable events	No. samples	Binomial prob.	Expected no.
0	5	0.0778	6.15
1	20	0.2592	20.48
2	26	0.3456	20.30
3	19	0.2304	18.20
4	7	0.0768	6.07
5	2	0.0102	0.81

No. observations/sample $= n = 5$
No. samples $= N = 79$

known to equal 0.4; thus, the probability of not observing a favorable outcome for any one observation equals $1 - p = 0.6$. To illustrate, the probability of occurrence of zero favorable events in samples of $n = 5$ observations is

$$P_{x=0} = \frac{5!}{5!(5-0)!}(0.4)^0(1-0.4)^{5-0}$$
$$= 0.0778$$

In order to compute the expected number of such outcomes for data consisting of $N = 79$ samples the probability of occurrence is multiplied by 79. For the case of zero favorable events one expects to observe $0.0778 \times 79 = 6.16$ samples. A chi square test of goodness of fit may now be applied. Here, the last two classes are combined and the number of restraints equals one, because of the use of n in calculating the expected values. This provides a total number of degrees of freedom equal to $5 - 1 = 4$. Thus, for the data of Table 10-2

$$\chi^2 = \frac{(5-6.15)^2}{6.15} + \cdots + \frac{[(7+2)-(6.07+0.81)]^2}{6.07+0.81}$$
$$= 0.980$$

The probability of observing a large χ^2 value lies between 0.90 and 0.95. Thus, one may safely accept the hypothesis that the observed data are binomially distributed.

It so happens that the true value of p was assumed known for these data, though most of the time one is not so fortunate. However, as might be expected, p can be estimated from the proportion of favorable events observed (p_{obs}) in all samples. Specifically, there are $5 \times 79 = 395$ events recorded in the N samples of which there are $0(5) + \cdots + 5(2) = 167$ favorable events. Thus

$$p_{obs} = 167/395 = 0.423$$

If we had used p_{obs} in setting up the binomial distribution goodness of fit, the distribution would have had a slightly different shape and the chi square value would have had only three degrees of freedom due to the extra restriction posed by the use of p_{obs} in computing the expected values.

Poisson distribution

This discrete distribution has not been discussed before. It is like the binomial in that it describes two-choice situations but only in situations where the likelihood of a favorable event is very small and the number of trials, by comparison, is very large. For example, if we had a binomial where, say, $p = 0.0001$ and $n = 500$, computing the expansion, $(p + q)^{n=500}$ would require the use of a computer. However, the Poisson distribution closely approximates the appropriate binomial in just these circumstances.

A Poisson probability of occurrence of exactly X favorable events, P_x, can be computed from the formula

$$P_x = \frac{\mu^x}{X!} e^{-\mu}$$

where e is the base of the natural logarithm and equals 2.718 . . . , and μ is the mean of the distribution of variables. Since μ is only rarely known it can be estimated by $\bar{X}$, the sample statistic. The value $e^{-\mu}$ is estimated by $e^{-\bar{x}}$ and is readily obtained in a table of exponentials found in most mathematics and physical science handbooks.

As the formula for the Poisson involves only one parameter, μ, it follows that its value determines the shape of the distribution. When μ is small the curve tends to be highly skewed toward the origin with a long tail, but as μ increases it tends to normality. The variance, σ^2, of the Poisson distribution is unique in that it can be found as

$$\sigma^2 = \mu$$

The values of n, the number of observations per sample, or p, the frequency of favorable events, are not immediately apparent in the probability formula for this distribution. However, they can be estimated as the product, np_{obs}, since

$$np_{\text{obs}} = \bar{X}$$

To illustrate this distribution, examine the data of Table 10-3 where are shown the results of a test of a new insecticide specific for a certain type of disease-carrying insect. The chemical was sprayed over several square miles of swampy area to test its effectiveness. The treated area was then sampled by means of 200 two-feet-square grids and the

Table 10-3 Observed number of surviving insects per grid fitted to a Poisson distribution

No. insects/grid	No. grids	Poisson prob.	Expected no.
0	106	0.507	101.40
1	63	0.345	69.00
2	23	0.117	23.40
3	6	0.027	5.40
4	1	0.004	0.80
5 or more	1	0.001	0.20
$N = 200$			

number of surviving insects of this type in each was recorded. One would expect some of the samples (grids) to show more survivors than others, but the question here is: Are the numbers of survivors randomly distributed or are there locales in the swamp where the insects, for one reason or another, are resistant to the insecticide?

The two choices here are: Presence or Absence of insects in a sample, letting "presence" represent the statistically favorable event, i.e., the event of prime interest.

First, one must estimate μ as

$$\begin{aligned}\bar{X} = np_{\text{obs}} &= 136 \text{ insects/200 grids}\\ &= 0.680 \text{ insect/grid}\end{aligned}$$

The Poisson probabilities for exactly X favorable events are computed according to the formula noted earlier. To illustrate, the probability of zero favorable events is computed as

$$P_{x=0} = \frac{0.680}{0!}\, 2.718^{-0.680} = \frac{1}{1}(0.507) = 0.507$$

The expected numbers are computed in the usual way: each Poisson probability is multiplied by the total number of samples ($N = 200$ in this example).

A chi square test of goodness of fit can now be computed. By convention the last two classes—4, and 5 or more—are lumped together so that

$$\begin{aligned}\chi^2 &= \frac{(106 - 101.40)^2}{101.40} + \cdots + \frac{[(1 + 1) - (0.80 + 0.20)]^2}{0.80 + 0.20}\\ &= 1.347\end{aligned}$$

After lumping, there are five classes remaining and one restriction due to μ being estimated by $\bar{X}$. Thus, there are $5 - 1 = 4$ degrees of freedom. The probability that the observed data were drawn from a Poisson distributed population lies in the interval $0.95 < P < 0.90$. Thus, one can readily accept the hypothesis that the surviving insects are randomly distributed and do not represent local pockets of resistance.

Tests of relationships

The remainder of this chapter will deal with situations where sets of variables are discontinuous or the data obtained are so incomplete that no standard test of relationship (e.g., correlation or regression) can be applied and one can determine only whether the variables are independent or, conversely, related to one another.

Consider the case where one is faced with the treatment of data that are classified by variables each of which contains but a very few categories, or where one variable has many categories but the others have few, and so on.

The same general procedure can be applied to all such *contingency* analyses. For example, one of the simplest and most common such tests of relationship is that in which each set of variables has but two categories which are then arrayed in a 2×2 contingency table, although this pattern can readily be extended to $r \times c$ tables where r and c denote numbers of categories in the rows and columns of the table. The hypothesis that is always tested first in these instances is the one of *independence of the variables*—that is, lack of relationship—and it is only after the data have been shown to lack this quality that one can assume a relationship between them.

In the data of Table 10-4A we see a 3×2 (that is, six-celled) contingency classification denoting the relationships of blood sugar curves observed in 30 pairs of identical and 32 pairs of nonidentical twins. Here, O_{ij} represents the observed value in the cell of the ith row and jth column, so that this example has $r = 1, \ldots, i = 2$ and $c = 1, \ldots, j = 3$. The hypothesis of independence to be tested here is: There is no relationship between the type of blood sugar curve and the type of twin sibs compared.

At this point it must be determined what values are expected when the two variables, twin type and blood sugar curve, are independent since it is to these expectations of independence that the observed

Table 10-4 Comparisons of blood sugar curves among identical and nonidentical twin sibs

	Twin type	Very similar	Intermediate	Very dissimilar	Total
A	Identical	$O_{11} = 10$	$O_{12} = 14$	$O_{13} = 6$	$O_{1.} = 30$
	Nonidentical	$O_{21} = 3$	$O_{22} = 9$	$O_{23} = 20$	$O_{2.} = 32$
	Totals	$O_{.1} = 13$	$O_{.2} = 23$	$O_{.3} = 26$	$O_{..} = 62$
B	Identical	$E_{11} = 6.3$	$E_{12} = 11.1$	$E_{13} = 12.6$	
	Nonidentical	$E_{21} = 6.7$	$E_{22} = 11.9$	$E_{23} = 13.4$	

values will be compared. Each E_{ij} in Table 10-4B represents the expected value for that cell when the variables are independent. The expected value for a cell is calculated from the product of the observed total for the row, $O_{..}$, and the column, $O_{.j}$, in which it appears divided by the grand total, $O_{i.}$. Thus, the general formula for computing an expected value for a cell is

$$E_{ij} = \frac{O_{i.}\, O_{.j}}{O_{..}}$$

and this formula works equally well for all degrees of $r \times c$ contingency classifications. Thus, for cell_{11}

$$E_{11} = \frac{30 \times 13}{62} = 6.3$$

The remaining expected values are computed in similar fashion.

Partial χ^2 values—that is, $(\text{Obs} - \text{Exp})^2/\text{Exp}$—have been computed for each of the cells of the table and are shown in Table 10-5. At this juncture it is evident from nothing more than a brief inspection of these values that the observations and expectations do not agree

Table 10-5 Partial χ^2 values computed for the data of Table 10-4

$\chi^2_{11} = 2.173$	$\chi^2_{12} = 0.757$	$\chi^2_{13} = 3.457$
$\chi^2_{21} = 2.043$	$\chi^2_{22} = 0.707$	$\chi^2_{23} = 3.251$

particularly well. However, it is not prudent to make a judgment on goodness of fit until the complete (that is, usual) chi square test is run. Here, the *complete* χ^2 value is equal to the sum of the partial χ^2 values, so that

$$\chi^2 = 2.173 + .757 + \cdots + 3.251 = 12.388$$

with $(r - 1)(c - 1) = 2$ degrees of freedom.

While this test is called chi square it is distributed only approximately as that quantity although it is close enough to be quite acceptable for the majority of such computations, provided that in one degree of freedom comparisons (e.g., 2×2 tables) the sample size is not "too small." (In a later portion of this chapter we will discuss the restrictions imposed by sample size in more detail.)

As can be seen from an examination of Table C, the probability associated with this chi square value is <0.005 so that there is little reason to hesitate in rejecting the hypothesis of independence. Clearly, the type of twin sibs compared and their respective blood sugar curves give strong indication of being related.

A useful procedure for determining the cause of significance if it is judged to be present is to eliminate the entire row(s) or column(s) which appear responsible for the significant χ^2 value, and then to run a test of significance on the abbreviated data. For the data of this experiment this procedure hardly seems necessary since the partial χ^2 values for the majority of the cells indicate very poor agreement between expected and observed values.

After concluding that contingency data are related, the obvious question that now arises is: What is the *intensity* of the relationship? The contingency classification is used in place of the correlation analysis when there are insufficient data to apply to the latter; however, it is a rather imperfect substitute since the correlation analysis immediately quantifies the relationship by a value of r. A statistic readily available in the contingency classification is the χ^2; intuitively, it would seem permissible to regard large χ^2 values as indications of greater intensities of relationship than smaller ones. Generally speaking, however, χ^2 values cannot be interpreted in this way since they are sensitive to *absolute differences* between observed and expected values. For example, the total χ^2 value for the data of Table 10-4 equals 12.388 with 2 degrees of freedom. If the sample size were doubled in each cell of that table so that $O_{..}$ now equaled 124 instead of the previous value of 62, the new χ^2 value would also be doubled—that is, 24.776 with 2 degrees of freedom. Thus, even though the χ^2 value is now twice as large as it was previously, the intensity of the relationship between the variables has not changed in the slightest degree. Then, what sort of solution to this problem is there? No concrete one, really, since no one has worked out a satisfactory test for measuring intensity of relationship in a contingency classification. However, conceivably one can work out one's own measure based on the observations themselves. For example, the proportion of the observed total contributed by a cell, per unit of the expected proportion attributable to that cell. That is

$$\frac{O_{ij}/O_{..}}{E_{ij}/O_{..}} = \frac{O_{ij}}{E_{ij}}$$

Another comparison might be

$$\frac{O_{ij} - E_{ij}}{E_{ij}/O_{..}}$$

where the absolute deviation between observed and expected values is measured per unit of contribution made by that cell under the condition of independence between variables. Regardless, the choices of test are varied and depend primarily on the hypothesis set forth regarding the nature of the relationship between the variables.

"Small" samples in chi square analyses often provide misleading conclusions as they are only approximately distributed as this quantity. One widely used method of circumventing this problem in 2 × 2 contingency tables—the only type of table where it can be applied—where the problem is especially prevalent due to their usually small sample size is called *Yates' correction for single degree of freedom comparisons* and it uses a modified form of the chi square test. It should be cautioned, however, that Yates' correction can further distort already biased data unless it is judiciously applied. While there is no general agreement as to the requirements for its application, Simpson et al.† suggest that it be applied only where the grand total, $O_{..}$, exceeds 40 and the number of observations in the smallest cell is equal to or less than 10. In the event that each cell has more than 10 observations the usual (that is, uncorrected) χ^2 should be applied as it works quite well, as it lso adoes in those cases where $O_{..} \leq 40$ and each cell has more than 10 observations. In this last instance it is best to calculate both Yates, adjusted and the unadjusted chi square tests. If both agree, then the hypothesis of independence is accepted or rejected depending on their mode of agreement. However, if the adjusted χ^2 value suggests independence while the unadjusted suggests the opposite, most experimenters generally find it reasonably satisfactory to accept the hypothesis of independence provided that they are also able to defend it on other—that is, biological—grounds.

Yates' adjusted chi square is found as

$$\chi^2 = \sum \frac{(|\text{Obs} - \text{Exp}| - 0.5)^2}{\text{Exp}}$$

† Simpson, Roe, and Lewontin (New York: Harcourt Brace, Jovanovich, 190).

with $(r - 1)(c - 1)$ degrees of freedom. Here, $|\text{Obs} - \text{Exp}|$ denotes "absolute value"; that is, the difference is always treated as positive. This correction consists of subtracting 0.5 from the absolute differences such that the value is always reduced by one-half unit toward zero.

Sometimes a different alternative to increasing numbers of observations in cells provides satisfactory for $r \times c$ tables of the size 2×2 and larger. It requires the combining of observations in adjacent rows or columns. This route should always be followed with a certain amount of hesitancy since it is often those cells which deviate most from the expectations which contain some of the most interesting information. Since these tables must always be symmetrical, cells cannot be combined at random; instead, entire rows or columns must be involved to maintain the symmetry. For example, if this procedure is applied to a 4×5 table, it may become either a 5×3 or a 4×4. Even if there is no a priori reason for choosing one format over the other, the experimenter should base his decision on those combinations of rows and columns which provide him with the maximum degrees of freedom after combining. By this criterion, then, the 4×4 adjustment is to be preferred over the 5×3 since the former provides 9 and the latter 8 degrees of freedom.

Homogeneity

Early in this book we began our discussions of comparisons of samples classified by one or more criteria, or variables. With sufficient data any number of different tests of comparison can be applied to these samples, always with the aim of answering the question: Are the samples so similar—as regards mean, variance, or whatever—that one can conclude that the distributions of the variables in question are identical in them and in the parent populations?

A test of homogeneity in the present context, however, is not so specific as to permit comparisons of means or variances. However, tests do permit one to declare samples as being either "the same" (homogeneous) or "different" (heterogeneous) regarding the distributions of the variables in each of them and, of course, in the parent populations. One might ask, for example, if the sex ratio at the time of birth is homogeneous in several samples of newborn drawn from several locales. To test for homogeneity of the variable, sex ratio, in this example and in others that are similar one simply applies a straightforward $r \times c$ contingency classification to the data with one change: here the hypothesis of independence is interpreted to mean that there is no relationship between the variables in the samples and, so, the samples must all have the same distributions.

$r \times c \times n$ Classifications

When one considers variables of classification beyond those of the $r \times c$ contingency table he must abandon the relatively straightforward techniques common to two-dimensional arrays; now the problems will deal with more complex situations of relationships, homogeneities, or both.

There are several approaches to solving three (or more)-way tables by the use of tests that differ markedly from those with which we have become familiar. However, the approach to be described here, though not the simplest to apply, does utilize successive applications of $r \times c$ tables and the chi square goodness of fit test. Table 10-6 presents data

Table 10-6[a]

			Response to radiation		
	Strain	Sex	Sick	Well	Total
A	One & Two	M	101	23	124
		F	81	43	124
			182	66	248
B	One	M	61	11	72
		F	45	27	72
			106	38	144
C	Two	M	40	12	52
		F	36	16	52
			76	28	104

[a] Data drawn from individuals classified by their resistance to radiation sickness: a fixed period of time after being given a fixed dose of ionizing radiation, according to strain type and sex.

classified in three ways, where the tests of comparison will be both of relationship and homogeneity. Here, individuals were scored according to their resistance to radiation sickness (being classified as either well or sick) a fixed period of time after being given a fixed dose of ionizing radiation, according to the genetic lineage (Strain One or Two), and their sex (male or female). Table 10-6A shows the pooled data while 10-6B and 10-6C show the individual results within the two strains. If the proportions of males and females were the same in both the sick and well categories then it would be readily apparent that there is no relationship between sex and resistance to radiation sickness. However, since this is not the case, an analysis by sex is strongly indicated. The first approach might be to set up a straightforward 2 × 2 contingency table based on the data of Table 10-6A in which the observations from both strains are pooled. However, before proceeding it would be wise to reexamine the individual strain data of 10-6B and 10-6C. Here one can note (a) the obvious differences between the two strains in the ratio of sick : well males such that Strain One *seems* to show a stronger relationship between these variables than Strain Two and, (b) while each of the strains has an equal number of males and females the size of the sample in Strain One is nearly 40% larger than that of Strain Two ($N = 144$ vs $N = 104$).

As was noted earlier, χ^2 is strongly influenced by sample size. This discrepancy alone should cause one to treat the two strains separately; however, this fact coupled with the one noted earlier regarding the possible strain differences in strength of relationship between sex and resistance makes the separate analyses of the strains mandatory. This has been done as shown in Table 10-7A and 10-7B where the expected values are shown, in parentheses, just below the observed. In Strain One it appears that sex and resistance are related ($P < 0.005$). In Strain Two, however, one sees just the opposite result, as the probability of obtaining a χ^2 value larger than the one observed lies between 0.25 and 0.50.

As was noted in the data of Table 10-4, χ^2's are additive provided they are independent (that is, contain independent observations) so that one can pool the χ^2 value for Strains One and Two. This step permits us to pool the data from both strains without the bias inherent in the χ^2 values due to their dissimilar sample sizes; both χ^2's are now weighted equally. Here, the *pooled* χ^2 is equal to

$$\chi_p^2 = \chi_1^2 + \chi_2^2 = 9.152 + 0.782 = 9.934$$

with $1 + 1 = 2$ pooled degrees of freedom. The probability associated with this new χ^2 values is <0.005. Taken at face value this new probability would seem to reaffirm the notion that the sexes do, indeed, demonstrate differential resistance to radiation sickness. However, there still arises one problem: The χ^2 value for Strain One contributed

Table 10-7 Contingency classification for a portion of the data arrayed in Table 10-6

Strain One			
	M	6 (53.0)	11 (19.0)
	F	45 (53.0)	27 (19.0)
$\chi^2 = 9.152$ with $df = 2$ $P < 0.005$			
Strain Two			
	M	40 (38.0)	12 (14.0)
	F	36 (38.0)	16 (14.0)
$\chi^2 = 0.782$ with $df = 2$ $0.50 < P < 0.25$			

so heavily to the pooled value as to virtually guarantee a significant relationship by itself—even with double its previous number of degrees of freedom.

How valid, then, is this new pooled value or, taken another way, how can one interpret the bias inherent in this value due to dissimilar strengths of relationship? One point that strongly favors the acceptance of this pooled χ^2 value, however, is the observation that both strains show the same *trend* in results, albeit to apparently differing degrees. For example, consider the frequencies of sick males and females in both strains.

	Sick		
	Males	Females	Male/Female
Strain One	0.833	0.625	1.333
Strain Two	0.769	0.692	1.111

Both strains have more than the expected frequency of sick males though Strain One has a ratio of sick males : females some 20% greater than that of Strain Two.

There is still one more test of confirmation that can be applied to these pooled data. It is based on the fact that

$$z = \sqrt{\frac{\sum \chi^2}{df}} = \frac{\sqrt{\chi_1^2} + \sqrt{\chi_2^2}}{\sqrt{\text{pooled } df}} = \frac{\chi_1 + \chi_2}{\sqrt{\text{pooled } df}}$$

For the data of Table 10-7A and 10-7B

$$z = \frac{\sqrt{9.152} + \sqrt{0.782}}{\sqrt{1+1}} = \frac{\pm 3.025 + \pm 0.884}{\pm 1.414} = \pm 2.77$$

Had the trends in the two strains been in opposite directions, one would have assigned opposite signs to the two χ values in the numerator and added them, algebraically, before dividing by degrees of freedom. Since the sign of the value $\sqrt{\chi^2}$ can be $\pm$ we will determine the probability of a z value lying in the interval ± 2.77. Table A in the Appendix shows that this probability is equal to $1 - 2(0.0028) = 0.9944$. The probability of falling outside this interval is so small that one is left with only one reasonable conclusion: The sexes do respond differently in their resistance to radiation sickness.

Finally, there remains the test of homogeneity: Are the distributions of the variables the same in both strains? The visual inspection of these two samples suggested a difference. However, this must be given closer scrutiny. One can set up a 4 × 2 table with males sick, males well, females sick, and females well, for each of the two strains since the proportions of males and females and sick and well individuals have real significance and are of equal importance in the context of this experiment. When this is done as shown in Table 10-8 one sees that the two strains appear homogeneous ($0.50 < P < 0.25$) as regards their resistance to radiation sickness.

Table 10-8 Contingency classification for a portion of the data shown in Table 10-6

Strain	M Sick	M Well	F Sick	F Well	Total
One	61	11	45	27	144
	(58.6)	(13.4)	(47.0)	(25.0)	
Two	40	12	36	16	104
	(42.4)	(9.6)	(34.0)	(18.0)	
Total	101	23	81	43	248

$\chi^2 = 1.849$ with $df = 3$
$0.50 < P < 0.25$

PROBLEMS

1. A parasitologist examined the liver of an individual who had been afflicted with a parasitic infection. Samples of liver tissue were drawn at random and the parasites (per cm³ of tissue) were counted.

No. parasites/cm^3	Obs. frequency
0	306
1	75
2	22
3	12
4	2
5	1
6 or more	0

(a) What type of distribution do the observations seem to fit? (b) Calculate the expected numbers utilizing the assumed frequency distribution. (c) Compute the goodness of fit of the expected and observed frequency distributions. (d) Test the hypothesis that the protozoa are randomly distributed within the liver.

2.

Treatment	Blooms	No blooms
1	10	30
2	9	31
3	9	41
4	13	27

Determine whether the ratio of blooming to nonblooming plants differs from treatment to treatment by more than chance.

3. An inspector in a ball bearing factory took 5-pound samples of bearings from a large company stockpile and counted the number of bearings in each sample that bore flaws.

Defective bearings per sample	Obs. frequency
0	5
1	13
2	23
3	21
4	17
5	11
6	5
7	2
8	1
9	0

Test the hypothesis that the flawed bearings occur randomly in the stockpile.

4. In the following data

Group	Obs. frequency
1	926
2	293
3	285
4	107

determine whether the observations fit a 9 : 3 : 3 : 1 ratio.

5. The following data relate age and sex to resistance to infection among laboratory animals.

Age group	Sex	Resistant	Susceptible
I	male	15	5
	female	23	1
II	male	5	3
	female	8	3
III	male	3	5
	female	7	4
IV	male	55	6
	female	34	17

(a) Are the factors age, sex, resistance independent of one another? (b) Is resistance independent of age group? (c) Is resistance independent of sex? (d) What do you conclude about the relationship age vs sex?

ANSWERS TO PROBLEMS

1. (a) Poisson
   (b) 277.0, 114.0, 23.5, 3.2, 0.3, 0.0, 0.0
   (c) $\chi^2 = 20.6$
   (d) $P < 0.01$
2. $\chi^2 = 2.63$; ($P > 0.05$)
3. Expected freq.

4.9	10.2
14.8	5.1
22.3	2.2
22.4	0.8
16.9	0.3

4. Expected freq.

906.2
302.1
302.1
100.7

APPENDIX

Numerical table Squares, square roots, and reciprocals

N	N^2	$\sqrt{N}$	$1/N$	N^2	N	$\sqrt{N}$	$1/N$
1	1	1.0000	1.000000	31	961	5.5678	.032258
2	4	1.4142	.500000	32	1024	5.6569	.031250
3	9	1.7321	.333333	33	1089	5.7446	.030303
4	16	2.0000	.250000	34	1156	5.8310	.029412
5	25	2.2361	.200000	35	1225	5.9161	.028571
6	36	2.4495	.166667	36	1296	6.0000	.027778
7	49	2.6458	.142857	37	1369	6.0828	.027027
8	64	2.8284	.125000	38	1444	6.1644	.026316
9	81	3.0000	.111111	39	1521	6.2450	.025641
10	100	3.1623	.100000	40	1600	6.3246	.025000
11	121	3.3166	.090909	41	1681	6.4031	.024390
12	144	3.4641	.083333	42	1764	6.4807	.023810
13	169	3.6056	.076923	43	1849	6.5574	.023256
14	196	3.7417	.071429	44	1936	6.6332	.022727
15	225	3.8730	.066667	45	2025	6.7082	.022222
16	256	4.0000	.062500	46	2116	6.7823	.021739
17	289	4.1231	.058824	47	2209	6.8557	.021277
18	324	4.2426	.055556	48	2304	6.9282	.020833
19	361	4.3589	.052632	49	2401	7.0000	.020408
20	400	4.4721	.050000	50	2500	7.0711	.020000
21	441	4.5826	.047619	51	2601	7.1414	.019608
22	484	4.6904	.045455	52	2704	7.2111	.019231
23	529	4.7958	.043478	53	2809	7.2801	.018868
24	576	4.8990	.041667	54	2916	7.3485	.018519
25	625	5.0000	.040000	55	3025	7.4162	.018182
26	676	5.0990	.038462	56	3136	7.4833	.017857
27	729	5.1962	.037037	57	3249	7.5498	.017544
28	784	5.2915	.035714	58	3364	7.6158	.017241
29	841	5.3852	.034483	59	3481	7.6811	.016949
30	900	5.4772	.033333	60	3600	7.7460	.016667

SOURCE: Portions of this table have been reproduced from *Handbook of Chemistry and Physics*, 52nd ed., 1971–72, Chemical Rubber Company, Cleveland, Ohio, by permission of the editors and publishers.

Numerical table (*continued*)

N	N^2	$\sqrt{N}$	$1/N$	N	N^2	$\sqrt{N}$	$1/N$
61	3721	7.8102	.016393	96	9216	9.7980	.010417
62	3844	7.8740	.016129	97	9409	9.8489	.010309
63	3969	7.9373	.015873	98	9604	9.8995	.010204
64	4096	8.0000	.015625	99	9801	9.9499	.010101
65	4225	8.0623	.015385	100	10000	10.0000	.010000
66	4356	8.1240	.015152	101	10201	10.0499	.00990099
67	4489	8.1854	.014925	102	10404	10.0995	.00980392
68	4624	8.2462	.014706	103	10609	10.1489	.00970874
69	4761	8.3066	.014493	104	10816	10.1980	.00961538
70	4900	8.3666	014286	105	11025	10.2470	.00952381
71	5041	8.4261	.014085	106	11236	10.2956	.00943396
72	5184	8.4853	.013889	107	11449	10.3441	.00934579
73	5329	8.5440	.013699	108	11664	10.3923	.00925926
74	5476	8.6023	.013514	109	11881	10.4403	.00917431
75	5625	8.6603	.013333	110	12100	10.4881	.00909091
76	5776	8.7178	.013158	111	12321	10.5357	.00900901
77	5929	8.7750	.012987	112	12544	10.5830	.00892857
78	6084	8.8318	.012821	113	12769	10.6301	.00884956
79	6241	8.8882	.012658	114	12996	10.6771	.00877193
80	6400	8.9443	.012500	115	13225	10.7238	.00869565
81	6561	9.0000	.012346	116	13456	10.7703	.00862069
82	6724	9.0554	.012195	117	13689	10.8167	.00854701
83	6889	9.1104	.012048	118	13924	10.8628	.00847458
84	7056	9.1652	.011905	119	14161	10.9087	.00840336
85	7225	9.2195	.011765	120	14400	10.9545	.00833333
86	7396	9.2736	.011628	121	14641	11.0000	.00826446
87	7569	9.3274	.011494	122	14884	11.0454	.00819672
88	7744	9.3808	.011364	123	15129	11.0905	.00813008
89	7921	9.4340	.011236	124	15376	11.1355	.00806452
90	8100	9.4868	.011111	125	15625	11.1803	.00800000
91	8281	9.5394	.010989	126	15876	11.2250	.00793651
92	8464	9.5917	.010870	127	16129	11.2694	.00787402
93	8649	9.6437	.010753	128	16384	11.3137	.00781250
94	8836	9.6954	.010638	129	16641	11.3578	.00775194
95	9025	9.7468	.010526	130	16900	11.4018	.00769231

Numerical table (*continued*)

N	N^2	$\sqrt{N}$	$1/N$	N	N^2	$\sqrt{N}$	$1/N$
131	17161	11.4455	.00763359	166	27556	12.8841	.00602410
132	17424	11.4891	.00757576	167	27889	12.9228	.00598802
133	17689	11.5326	.00751880	168	28224	12.9615	.00595238
134	17956	11.5758	.00746269	169	28561	13.0000	.00591716
135	18225	11.6190	.00740741	170	28900	13.0384	.00588235
136	18496	11.6619	.00735294	171	29241	13.0767	.00584795
137	18769	11.7047	.00729927	172	29584	13.1149	.00581395
138	19044	11.7473	.00724638	173	29929	13.1529	.00578035
139	19321	11.7898	.00719424	174	30276	13.1909	.00574713
140	19600	11.8322	.00714286	175	30625	13.2288	.00571429
141	19881	11.8743	.00709220	176	30976	13.2665	.00568182
142	20164	11.9164	.00704225	177	31329	13.3041	.00564972
143	20449	11.9583	.00699301	178	31684	13.3417	.00561798
144	20736	12.0000	.00694444	179	32041	13.3791	.00558659
145	21025	12.0416	.00689655	180	32400	13.4164	.00555556
146	21316	12.0830	.00684932	181	32761	13.4536	.00552486
147	21609	12.1244	.00680272	182	33124	13.4907	.00549451
148	21904	12.1655	.00675676	183	33489	13.5277	.00546448
149	22201	12.2066	.00671141	184	33856	13.5647	.00543478
150	22500	12.2474	.00666667	185	34225	13.6015	.00540541
151	22801	12.2882	.00662252	186	34596	13.6382	.00537634
152	23104	12.3288	.00657895	187	34969	13.6748	.00534759
153	23409	12.3693	.00653595	188	35344	13.7113	.00531915
154	23716	12.4097	.00649351	189	35721	13.7477	.00529101
155	24025	12.4499	.00645161	190	36100	13.7840	.00526316
156	24336	12.4900	.00641026	191	36481	13.8203	.00523560
157	24649	12.5300	.00636943	192	36864	13.8564	.00520833
158	24964	12.5698	.00632911	193	37249	13.8924	.00518135
159	25281	12.6095	.00628931	194	37636	13.9284	.00515464
160	25600	12.6491	.00625000	195	38025	13.9642	.00512821
161	25921	12.6886	.00621118	196	38416	14.0000	.00510204
162	26244	12.7279	.00617284	197	38809	14.0357	.00507614
163	26569	12.7671	.00613497	198	39204	14.0712	.00505051
164	26896	12.8062	.00609756	199	39601	14.1067	.00502513
165	27225	12.8452	.00606061	200	40000	14.1421	.00500000

Table A

Probability of a random value of z being greater than the values shown

z	.00	.01	.02	.03	.04	.05	.06	.07	.08	.09
.0	.5000	.4960	.4920	.4880	.4840	.4801	.4761	.4721	.4681	.4641
.1	.4602	.4562	.4522	.4483	.4443	.4404	.4364	.4325	.4286	.4247
.2	.4207	.4168	.4129	.4090	.4052	.4013	.3974	.3936	.3897	.3859
.3	.3821	.3783	.3745	.3707	.3669	.3632	.3594	.3557	.3520	.3483
.4	.3446	.3409	.3372	.3336	.3300	.3264	.3228	.3192	.3156	.3121
.5	.3085	.3050	.3015	.2981	.2946	.2912	.2877	.2843	.2810	.2776
.6	.2743	.2709	.2676	.2643	.2611	.2578	.2546	.2514	.2483	.2451
.7	.2420	.2389	.2358	.2327	.2296	.2266	.2236	.2206	.2177	.2148
.8	.2119	.2090	.2061	.2033	.2005	.1977	.1949	.1922	.1894	.1867
.9	.1841	.1814	.1788	.1762	.1736	.1711	.1685	.1660	.1635	.1611
1.0	.1587	.1562	.1539	.1515	.1492	.1469	.1446	.1423	.1401	.1379
1.1	.1357	.1335	.1314	.1292	.1271	.1251	.1230	.1210	.1190	.1170
1.2	.1151	.1131	.1112	.1093	.1075	.1056	.1038	.1020	.1003	.0985
1.3	.0968	.0951	.0934	.0918	.0901	.0885	.0869	.0853	.0838	.0823
1.4	.0808	.0793	.0778	.0764	.0749	.0735	.0721	.0708	.0694	.0681
1.5	.0668	.0655	.0643	.0630	.0618	.0606	.0594	.0582	.0571	.0559
1.6	.0548	.0537	.0526	.0516	.0505	.0495	.0485	.0475	.0465	.0455
1.7	.0446	.0436	.0427	.0418	.0409	.0401	.0392	.0384	.0375	.0367
1.8	.0359	.0351	.0344	.0336	.0329	.0322	.0314	.0307	.0301	.0294
1.9	.0287	.0281	.0274	.0268	.0262	.0256	.0250	.0244	.0239	.0233
2.0	.0228	.0222	.0217	.0212	.0207	.0202	.0197	.0192	.0188	.0183
2.1	.0179	.0174	.0170	.0166	.0162	.0158	.0154	.0150	.0146	.0143
2.2	.0139	.0136	.0132	.0129	.0125	.0122	.0119	.0116	.0113	.0110
2.3	.0107	.0104	.0102	.0099	.0096	.0094	.0091	.0089	.0087	.0084
2.4	.0082	.0080	.0078	.0075	.0073	.0071	.0069	.0068	.0066	.0064
2.5	.0062	.0060	.0059	.0057	.0055	.0054	.0052	.0051	.0049	.0048
2.6	.0047	.0045	.0044	.0043	.0041	.0040	.0039	.0038	.0037	.0036
2.7	.0035	.0034	.0033	.0032	.0031	.0030	.0029	.0028	.0027	.0026
2.8	.0026	.0025	.0024	.0023	.0023	.0022	.0021	.0021	.0020	.0019
2.9	.0019	.0018	.0018	.0017	.0016	.0016	.0015	.0015	.0014	.0014
3.0	.0013	.0013	.0013	.0012	.0012	.0011	.0011	.0011	.0010	.0010
3.1	.0010	.0009	.0009	.0009	.0008	.0008	.0008	.0008	.0007	.0007
3.2	.0007	.0007	.0006	.0006	.0006	.0006	.0006	.0005	.0005	.0005
3.3	.0005	.0005	.0005	.0004	.0004	.0004	.0004	.0004	.0004	.0003
3.4	.0003	.0003	.0003	.0003	.0003	.0003	.0003	.0003	.0003	.0002
3.6	.0002	.0002	.0001	.0001	.0001	.0001	.0001	.0001	.0001	.0001
3.9	.0000									

Table B Student's t distribution

Cumulative probability between $+t$ and $-t$

df	.1	.2	.3	.4	.5	.6	.7	.8	.9	.95	.98	.99	.999
1	.158	.325	.510	.727	1.000	1.376	1.963	3.078	6.314	12.706	31.821	63.657	636.619
2	.142	.289	.445	.617	.816	1.061	1.386	1.886	2.920	4.303	6.965	9.925	31.598
3	.137	.277	.424	.584	.765	.978	1.250	1.638	2.353	3.182	4.541	5.841	12.941
4	.134	.271	.414	.569	.741	.941	1.190	1.533	2.132	2.776	3.747	4.604	8.610
5	.132	.267	.408	.559	.727	.920	1.156	1.476	2.015	2.571	3.365	4.032	6.859
6	.131	.265	.404	.553	.718	.906	1.134	1.440	1.943	2.447	3.143	3.707	5.959
7	.130	.263	.402	.549	.711	.896	1.119	1.415	1.895	2.365	2.998	3.499	5.405
8	.130	.262	.399	.546	.706	.889	1.108	1.397	1.860	2.306	2.896	3.355	5.041
9	.129	.261	.398	.543	.703	.883	1.100	1.383	1.833	2.262	2.821	3.250	4.781
10	.129	.260	.397	.542	.700	.879	1.093	1.372	1.812	2.228	2.764	3.169	4.587
11	.129	.260	.396	.540	.697	.876	1.088	1.363	1.796	2.201	2.718	3.106	4.437
12	.128	.259	.395	.539	.695	.873	1.083	1.356	1.782	2.179	2.681	3.055	4.318
13	.128	.259	.394	.538	.694	.870	1.079	1.350	1.771	2.160	2.650	3.012	4.221
14	.128	.258	.393	.537	.692	.868	1.076	1.345	1.761	2.145	2.624	2.977	4.140
15	.128	.258	.393	.536	.691	.866	1.074	1.341	1.753	2.131	2.602	2.947	4.073
16	.128	.258	.392	.535	.690	.865	1.071	1.337	1.746	2.120	2.583	2.921	4.015
17	.128	.257	.392	.534	.689	.863	1.069	1.333	1.740	2.110	2.567	2.898	3.965
18	.127	.257	.392	.534	.688	.862	1.067	1.330	1.734	2.101	2.552	2.878	3.922
19	.127	.257	.391	.533	.688	.861	1.066	1.328	1.729	2.093	2.539	2.861	3.883
20	.127	.257	.391	.533	.687	.860	1.064	1.325	1.725	2.086	2.528	2.845	3.850
21	.127	.257	.391	.532	.686	.859	1.063	1.323	1.721	2.080	2.518	2.831	3.819
22	.127	.256	.390	.532	.686	.858	1.061	1.321	1.717	2.074	2.508	2.819	3.792
23	.127	.256	.390	.532	.685	.858	1.060	1.319	1.714	2.069	2.500	2.807	3.767
24	.127	.256	.390	.531	.685	.857	1.059	1.318	1.711	2.064	2.492	2.797	3.745
25	.127	.256	.390	.531	.684	.856	1.058	1.316	1.708	2.060	2.485	2.787	3.725
26	.127	.256	.390	.531	.684	.856	1.058	1.315	1.706	2.056	2.479	2.779	3.707
27	.127	.256	.389	.531	.684	.855	1.057	1.314	1.703	2.052	2.473	2.771	3.690
28	.127	.256	.389	.530	.683	.855	1.056	1.313	1.701	2.048	2.467	2.763	3.674
29	.127	.256	.389	.530	.683	.854	1.055	1.311	1.699	2.045	2.462	2.756	3.659
30	.127	.256	.389	.530	.683	.854	1.055	1.310	1.697	2.042	2.457	2.750	3.646
40	.126	.255	.388	.529	.681	.851	1.050	1.303	1.684	2.021	2.423	2.704	3.551
60	.126	.254	.387	.527	.679	.848	1.046	1.296	1.671	2.000	2.390	2.660	3.460
120	.126	.254	.386	.526	.677	.845	1.041	1.289	1.658	1.980	2.358	2.617	3.373
∞	.126	.253	.385	.524	.674	.842	1.036	1.282	1.645	1.960	2.326	2.576	3.291
"2-sided"	.9	.8	.7	.6	.5	.4	.3	.2	.1	.05	.02	.01	.001
"1-sided"	.45	.40	.35	.30	.25	.20	.15	.10	.05	.025	.010	.005	.0005

Significance probabilities for t-test

Appendix Tables are taken from Table III of Fisher and Yates: *Statistical Tables for Biological, Agricultural, and Medical Research*, published by Oliver & Boyd, Edinburgh, and by permission of the authors and publishers.

Table C Values of χ^2

	Probability of a larger value of χ^2 being obtained by chance alone											
df	.995	.990	.975	.950	.900	.750	.500	.250	.100	.050	.025	.010
1	$.0^4393$	$.0^3157$	$.0^3982$	$.0^2393$	.0158	.102	.455	1.32	2.71	3.84	5.02	6.63
2	.0100	.0201	.0506	.103	.211	.575	1.39	2.77	4.61	5.99	7.38	9.21
3	.0717	.115	.216	.352	.584	1.21	2.37	4.11	6.25	7.81	9.35	11.3
4	.207	.297	.484	.711	1.06	1.92	3.36	5.39	7.78	9.49	11.1	13.3
5	.412	.554	.831	1.15	1.61	2.67	4.35	6.63	9.24	11.1	12.8	15.1
6	.676	.872	1.24	1.64	2.20	3.45	5.35	7.84	10.6	12.6	14.4	16.8
7	.989	1.24	1.69	2.17	2.83	4.25	6.35	9.04	12.0	14.1	16.0	18.5
8	1.34	1.65	2.18	2.73	3.49	5.07	7.34	10.2	13.4	15.5	17.5	20.1
9	1.73	2.09	2.70	3.33	4.17	5.90	8.34	11.4	14.7	16.9	19.0	21.7
10	2.16	2.56	3.25	3.94	4.87	6.74	9.34	12.5	16.0	18.3	20.5	23.2
11	2.60	3.05	3.82	4.57	5.58	7.58	10.3	13.7	17.3	19.7	21.9	24.7
12	3.07	3.57	4.40	5.23	6.30	8.44	11.3	14.8	18.5	21.0	23.3	26.2
13	3.57	4.11	5.01	5.89	7.04	9.30	12.3	16.0	19.8	22.4	24.7	27.7
14	4.07	4.66	5.63	6.57	7.79	10.2	13.3	17.1	21.1	23.7	26.1	29.1
15	4.60	5.23	6.26	7.26	8.55	11.0	14.3	18.2	22.3	25.0	27.5	30.6
16	5.14	5.81	6.91	7.96	9.31	11.9	15.3	19.4	23.5	26.3	28.8	32.0
17	5.70	6.41	7.56	8.67	10.1	12.8	16.3	20.5	24.8	27.6	30.2	33.4
18	6.26	7.01	8.23	9.39	10.9	13.7	17.3	21.6	26.0	28.9	31.5	34.8
19	6.84	7.63	8.91	10.1	11.7	14.6	18.3	22.7	27.2	30.1	32.9	36.2
20	7.43	8.26	9.59	10.9	12.4	15.5	19.3	23.8	28.4	31.4	34.2	37.6
21	8.03	8.90	10.3	11.6	13.2	16.3	20.3	24.9	29.6	32.7	35.5	38.9
22	8.64	9.54	11.0	12.3	14.0	17.2	21.3	26.0	30.8	33.9	36.8	40.3
23	9.26	10.2	11.7	13.1	14.8	18.1	22.3	27.1	32.0	35.2	38.1	41.6
24	9.89	10.9	12.4	13.8	15.7	19.0	23.3	28.2	33.2	36.4	39.4	43.0
25	10.5	11.5	13.1	14.6	16.5	19.9	24.3	29.3	34.4	37.7	40.6	44.3
26	11.2	12.2	13.8	15.4	17.3	20.8	25.3	30.4	35.6	38.9	41.9	45.6
27	11.8	12.9	14.6	16.2	18.1	21.7	26.3	31.5	36.7	40.1	43.2	47.0
28	12.5	13.6	15.3	16.9	18.9	22.7	27.3	32.6	37.9	41.3	44.5	48.3
29	13.1	14.3	16.0	17.7	19.8	23.6	28.3	33.7	39.1	42.6	45.7	49.6
30	13.8	15.0	16.8	18.5	20.6	24.5	29.3	34.8	40.3	43.8	47.0	50.9
40	20.7	22.2	24.4	26.5	29.1	33.7	39.3	45.6	51.8	55.8	59.3	63.7
50	28.0	29.7	32.4	34.8	37.7	42.9	49.3	56.3	63.2	67.5	71.4	76.2
60	35.5	37.5	40.5	43.2	46.5	52.3	59.3	67.0	74.4	79.1	83.3	88.4

SOURCE: Abridged from "Table of percentage points of the χ^2 distribution," *Biometrika* **32**: 188–189 (1941), by Catherine M. Thompson. We are grateful to the author and editor for permission.

Table D Cumulative *F* Distribution

Denominator *df*	"1-sided" Tests	Numerator *df* 1	2	3	4	5	6	7	8	9	10	12	15	20	30	60	120	∞	"2-sided" Tests
1	.10	39.9	49.5	53.6	55.8	57.2	58.2	58.9	59.4	59.9	60.2	60.7	61.2	61.7	62.3	62.8	63.1	63.3	.20
	.05	161	200	216	225	230	234	237	239	241	242	244	246	248	250	252	253	254	.10
	.025	648	800	864	900	922	937	948	957	963	969	977	985	993	1000	1010	1010	1020	.05
	.01	4,050	5,000	5,400	5,620	5,760	5,860	5,930	5,980	6,020	6,060	6,110	6,160	6,210	6,260	6,310	6,340	6,370	.02
	.005	16,200	20,000	21,600	22,500	23,100	23,400	23,700	23,900	24,100	24,200	24,400	24,600	24,800	25,000	25,300	25,400	25,500	.01
2	.10	8.53	9.00	9.16	9.24	9.29	9.33	9.35	9.37	9.38	9.39	9.41	9.42	9.44	9.46	9.47	9.48	9.49	.20
	.05	18.5	19.0	19.2	19.2	19.3	19.3	19.4	19.4	19.4	19.4	19.4	19.4	19.5	19.5	19.5	19.5	19.5	.10
	.025	38.5	39.0	39.2	39.2	39.3	39.3	39.4	39.4	39.4	39.4	39.4	39.4	39.4	39.5	39.5	39.5	39.5	.05
	.01	98.5	99.0	99.2	99.2	99.3	99.3	99.4	99.4	99.4	99.4	99.4	99.4	99.4	99.5	99.5	99.5	99.5	.02
	.005	199	199	199	199	199	199	199	199	199	199	199	199	199	199	199	199	200	.01
3	.10	5.54	5.46	5.39	5.34	5.31	5.28	5.27	5.25	5.24	5.23	5.22	5.20	5.18	5.17	5.15	5.14	5.13	.20
	.05	10.1	9.55	9.28	9.12	9.01	8.94	8.89	8.85	8.81	8.79	8.74	8.70	8.66	8.62	8.57	8.55	8.53	.10
	.025	17.4	16.0	15.4	15.1	14.9	14.7	14.6	14.5	14.5	14.4	14.3	14.3	14.2	14.1	14.0	13.9	13.9	.05
	.01	34.1	30.8	29.5	28.7	28.2	27.9	27.7	27.5	27.3	27.2	27.1	26.9	26.7	26.5	26.3	26.2	26.1	.02
	.005	55.6	49.8	47.5	46.2	45.4	44.8	44.4	44.1	43.9	43.7	43.4	43.1	42.8	42.5	42.1	42.0	41.8	.01
4	.10	4.54	4.32	4.19	4.11	4.05	4.01	3.98	3.95	3.94	3.92	3.90	3.87	3.84	3.82	3.79	3.78	3.76	.20
	.05	7.71	6.94	6.59	6.39	6.26	6.16	6.09	6.04	6.00	5.96	5.91	5.86	5.80	5.75	5.69	5.66	5.63	.10
	.025	12.2	10.6	9.98	9.60	9.36	9.20	9.07	8.98	8.90	8.84	8.75	8.66	8.56	8.46	8.36	8.31	8.26	.05
	.01	21.2	18.0	16.7	16.0	15.5	15.2	15.0	14.8	14.7	14.5	14.4	14.2	14.0	13.8	13.7	13.6	13.5	.02
	.005	31.3	26.3	24.3	23.2	22.5	22.0	21.6	21.4	21.1	21.0	20.7	20.4	20.2	19.9	19.6	19.5	19.3	.01
5	.10	4.06	3.78	3.62	3.52	3.45	3.40	3.37	3.34	3.32	3.30	3.27	3.24	3.21	3.17	3.14	3.12	3.10	.20
	.05	6.61	5.79	5.41	5.19	5.05	4.95	4.88	4.82	4.77	4.74	4.68	4.62	4.56	4.50	4.43	4.40	4.37	.10
	.025	10.0	8.43	7.76	7.39	7.15	6.98	6.85	6.76	6.68	6.62	6.52	6.43	6.33	6.23	6.12	6.07	6.02	.05
	.01	16.3	13.3	12.1	11.4	11.0	10.7	10.5	10.3	10.2	10.1	9.89	9.72	9.55	9.38	9.20	9.11	9.02	.02
	.005	22.8	18.3	16.5	15.6	14.9	14.5	14.2	14.0	13.8	13.6	13.4	13.1	12.9	12.7	12.4	12.3	12.1	.01
6	.10	3.78	3.46	3.29	3.18	3.11	3.05	3.01	2.98	2.96	2.94	2.90	2.87	2.84	2.80	2.76	2.74	2.72	.20
	.05	5.99	5.14	4.76	4.53	4.39	4.28	4.21	4.15	4.10	4.06	4.00	3.94	3.87	3.81	3.74	3.70	3.67	.10
	.025	8.81	7.26	6.60	6.23	5.99	5.82	5.70	5.60	5.52	5.46	5.37	5.27	5.17	5.07	4.96	4.90	4.85	.05
	.01	13.7	10.9	9.78	9.15	8.75	8.47	8.26	8.10	7.98	7.87	7.72	7.56	7.40	7.23	7.06	6.97	6.88	.02
	.005	18.6	14.5	12.9	12.0	11.5	11.1	10.8	10.6	10.4	10.2	10.0	9.81	9.59	9.36	9.12	9.00	8.88	.01
7	.10	3.59	3.26	3.07	2.96	2.88	2.83	2.78	2.75	2.72	2.70	2.67	2.63	2.59	2.56	2.51	2.49	2.47	.20
	.05	5.59	4.74	4.35	4.12	3.97	3.87	3.79	3.73	3.68	3.64	3.57	3.51	3.44	3.38	3.30	3.27	3.23	.10
	.025	8.07	6.54	5.89	5.52	5.29	5.12	4.99	4.90	4.82	4.76	4.67	4.57	4.47	4.36	4.25	4.20	4.14	.05
	.01	12.2	9.55	8.45	7.85	7.46	7.19	6.99	6.84	6.72	6.62	6.47	6.31	6.16	5.99	5.82	5.74	5.65	.02
	.005	16.2	12.4	10.9	10.1	9.52	9.16	8.89	8.68	8.51	8.38	8.18	7.97	7.75	7.53	7.31	7.19	7.08	.01
8	.10	3.46	3.11	2.92	2.81	2.73	2.67	2.62	2.59	2.56	2.54	2.50	2.46	2.42	2.38	2.34	2.32	2.29	.20
	.05	5.32	4.46	4.07	3.84	3.69	3.58	3.50	3.44	3.39	3.35	3.28	3.22	3.15	3.08	3.01	2.97	2.93	.10
	.025	7.57	6.06	5.42	5.05	4.82	4.65	4.53	4.43	4.36	4.30	4.20	4.10	4.00	3.89	3.78	3.73	3.67	.05
	.01	11.3	8.65	7.59	7.01	6.63	6.37	6.18	6.03	5.91	5.81	5.67	5.52	5.36	5.20	5.03	4.95	4.86	.02
	.005	14.7	11.0	9.60	8.81	8.30	7.95	7.69	7.50	7.34	7.21	7.01	6.81	6.61	6.40	6.18	6.06	5.95	.01

Table D (*continued*)

Denominator *df*	"1-sided" Tests	Numerator *df* 1	2	3	4	5	6	7	8	9	10	12	15	20	30	60	120	∞	"2 sided" Tests
9	.10	3.36	3.01	2.81	2.69	2.61	2.55	2.51	2.47	2.44	2.42	2.38	2.34	2.30	2.25	2.21	2.18	2.16	.20
	.05	5.12	4.26	3.86	3.63	3.48	3.37	3.29	3.23	3.18	3.14	3.07	3.01	2.94	2.86	2.79	2.75	2.71	.10
	.025	7.21	5.71	5.08	4.72	4.48	4.32	4.20	4.10	4.03	3.96	3.87	3.77	3.67	3.56	3.45	3.39	3.33	.05
	.01	10.6	8.02	6.99	6.42	6.06	5.80	5.61	5.47	5.35	5.26	5.11	4.96	4.81	4.65	4.48	4.40	4.31	.02
	.005	13.6	10.1	8.72	7.96	7.47	7.13	6.88	6.69	6.54	6.42	6.23	6.03	5.83	5.62	5.41	5.30	5.19	.01
10	.10	3.28	2.92	2.73	2.61	2.52	2.46	2.41	2.38	2.35	2.32	2.28	2.24	2.20	2.16	2.11	2.08	2.06	.20
	.05	4.96	4.10	3.71	3.48	3.33	3.22	3.14	3.07	3.02	2.98	2.91	2.84	2.77	2.70	2.62	2.58	2.54	.10
	.025	6.94	5.46	4.83	4.47	4.24	4.07	3.95	3.85	3.78	3.72	3.62	3.52	3.42	3.31	3.20	3.14	3.08	.05
	.01	10.0	7.56	6.55	5.99	5.64	5.39	5.20	5.06	4.94	4.85	4.71	4.56	4.41	4.25	4.08	4.00	3.91	.02
	.005	12.8	9.43	8.08	7.34	6.87	6.54	6.30	6.12	5.97	5.85	5.66	5.47	5.27	5.07	4.86	4.75	4.64	.01
12	.10	3.18	2.81	2.61	2.48	2.39	2.33	2.28	2.24	2.21	2.19	2.15	2.10	2.06	2.01	1.96	1.93	1.90	.20
	.05	4.75	3.89	3.49	3.26	3.11	3.00	2.91	2.85	2.80	2.75	2.69	2.62	2.54	2.47	2.38	2.34	2.30	.10
	.025	6.55	5.10	4.47	4.12	3.89	3.73	3.61	3.51	3.44	3.37	3.28	3.18	3.07	2.96	2.85	2.79	2.72	.05
	.01	9.33	6.93	5.95	5.41	5.06	4.82	4.64	4.50	4.39	4.30	4.16	4.01	3.86	3.70	3.54	3.45	3.36	.02
	.005	11.8	8.51	7.23	6.52	6.07	5.76	5.52	5.35	5.20	5.09	4.91	4.72	4.53	4.33	4.12	4.01	3.90	.01
15	.10	3.07	2.70	2.49	2.36	2.27	2.21	2.16	2.12	2.09	2.06	2.02	1.97	1.92	1.87	1.82	1.79	1.76	.20
	.05	4.54	3.68	3.29	3.06	2.90	2.79	2.71	2.64	2.59	2.54	2.48	2.40	2.33	2.25	2.16	2.11	2.07	.10
	.025	6.20	4.77	4.15	3.80	3.58	3.41	3.29	3.20	3.12	3.06	2.96	2.86	2.76	2.64	2.52	2.46	2.40	.05
	.01	8.68	6.36	5.42	4.89	4.56	4.32	4.14	4.00	3.89	3.80	3.67	3.52	3.37	3.21	3.05	2.96	2.87	.02
	.005	10.8	7.70	6.48	5.80	5.37	5.07	4.85	4.67	4.54	4.42	4.25	4.07	3.88	3.69	3.48	3.37	3.26	.01
20	.10	2.97	2.59	2.38	2.25	2.16	2.09	2.04	2.00	1.96	1.94	1.89	1.84	1.79	1.74	1.68	1.64	1.61	.20
	.05	4.35	3.49	3.10	2.87	2.71	2.60	2.51	2.45	2.39	2.35	2.28	2.20	2.12	2.04	1.95	1.90	1.84	.10
	.025	5.87	4.46	3.86	3.51	3.29	3.13	3.01	2.91	2.84	2.77	2.68	2.57	2.46	2.35	2.22	2.16	2.09	.05
	.01	8.10	5.85	4.94	4.43	4.10	3.87	3.70	3.56	3.46	3.37	3.23	3.09	2.94	2.78	2.61	2.52	2.42	.02
	.005	9.94	6.99	5.82	5.17	4.76	4.47	4.26	4.09	3.96	3.85	3.68	3.50	3.32	3.12	2.92	2.81	2.69	.01
30	.10	2.88	2.49	2.28	2.14	2.05	1.98	1.93	1.88	1.85	1.82	1.77	1.72	1.67	1.61	1.54	1.50	1.46	.20
	.05	4.17	3.32	2.92	2.69	2.53	2.42	2.33	2.27	2.21	2.16	2.09	2.01	1.93	1.84	1.74	1.68	1.62	.10
	.025	5.57	4.18	3.59	3.25	3.03	2.87	2.75	2.65	2.57	2.51	2.41	2.31	2.20	2.07	1.94	1.87	1.79	.05
	.01	7.56	5.39	4.51	4.02	3.70	3.47	3.30	3.17	3.07	2.98	2.84	2.70	2.55	2.39	2.21	2.11	2.01	.02
	.005	9.18	6.35	5.24	4.62	4.23	3.95	3.74	3.58	3.45	3.34	3.18	3.01	2.82	2.63	2.42	2.30	2.18	.01
60	.10	2.79	2.39	2.18	2.04	1.95	1.87	1.82	1.77	1.74	1.71	1.66	1.60	1.54	1.48	1.40	1.35	1.29	.20
	.05	4.00	3.15	2.76	2.53	2.37	2.25	2.17	2.10	2.04	1.99	1.92	1.84	1.75	1.65	1.53	1.47	1.39	.10
	.025	5.29	3.93	3.34	3.01	2.79	2.63	2.51	2.41	2.33	2.27	2.17	2.06	1.94	1.82	1.67	1.58	1.48	.05
	.01	7.08	4.98	4.13	3.65	3.34	3.12	2.95	2.82	2.72	2.63	2.50	2.35	2.20	2.03	1.84	1.73	1.60	.02
	.005	8.49	5.80	4.73	4.14	3.76	3.49	3.29	3.13	3.01	2.90	2.74	2.57	2.39	2.19	1.96	1.83	1.69	.01
120	.10	2.75	2.35	2.13	1.99	1.90	1.82	1.77	1.72	1.68	1.65	1.60	1.54	1.48	1.41	1.32	1.26	1.19	.20
	.05	3.92	3.07	2.68	2.45	2.29	2.18	2.09	2.02	1.96	1.91	1.83	1.75	1.66	1.55	1.43	1.35	1.25	.10
	.025	5.15	3.80	3.23	2.89	2.67	2.52	2.39	2.30	2.22	2.16	2.05	1.94	1.82	1.69	1.53	1.43	1.31	.05
	.01	6.85	4.79	3.95	3.48	3.17	2.96	2.79	2.66	2.56	2.47	2.34	2.19	2.03	1.86	1.66	1.53	1.38	.02
	.005	8.18	5.54	4.50	3.92	3.55	3.28	3.09	2.93	2.81	2.71	2.54	2.37	2.19	1.98	1.75	1.61	1.43	.01
∞	.10	2.71	2.30	2.08	1.94	1.85	1.77	1.72	1.67	1.63	1.60	1.55	1.49	1.42	1.34	1.24	1.17	1.00	.20
	.05	3.84	3.00	2.60	2.37	2.21	2.10	2.01	1.94	1.88	1.83	1.75	1.67	1.57	1.46	1.32	1.22	1.00	.10
	.025	5.02	3.69	3.12	2.79	2.57	2.41	2.29	2.19	2.11	2.05	1.94	1.83	1.71	1.57	1.39	1.27	1.00	.05
	.01	6.63	4.61	3.78	3.32	3.02	2.80	2.64	2.51	2.41	2.32	2.18	2.04	1.88	1.70	1.47	1.32	1.00	.02
	.005	7.88	5.30	4.28	3.72	3.35	3.09	2.90	2.74	2.62	2.52	2.36	2.19	2.00	1.79	1.53	1.36	1.00	.01

SOURCE: Adapted from Table 18 of *Biometrika Tables for Statisticians* (Volume I), by E. S. Pearson and H. O. Hartley, Cambridge University Press for the Biometrika Trustees.

Table E Critical values for the 5% level of Duncan's test

Error *df*	Level	Number of means in range being tested													
		2	3	4	5	6	7	8	9	10	12	14	16	18	20
1	.05	18.0	18.0	18.0	18.0	18.0	18.0	18.0	18.0	18.0	18.0	18.0	18.0	18.0	18.0
	.01	90.0	90.0	90.0	90.0	90.0	90.0	90.0	90.0	90.0	90.0	90.0	90.0	90.0	90.0
2	.05	6.09	6.09	6.09	6.09	6.09	6.09	6.09	6.09	6.09	6.09	6.09	6.09	6.09	6.09
	.01	14.0	14.0	14.0	14.0	14.0	14.0	14.0	14.0	14.0	14.0	14.0	14.0	14.0	14.0
3	.05	4.50	4.50	4.50	4.50	4.50	4.50	4.50	4.50	4.50	4.50	4.50	4.50	4.50	4.50
	.01	8.26	8.5	8.6	8.7	8.8	8.9	8.9	9.0	9.0	9.0	9.1	9.2	9.3	9.3
4	.05	3.93	4.01	4.02	4.02	4.02	4.02	4.02	4.02	4.02	4.02	4.02	4.02	4.02	4.02
	.01	6.51	6.8	6.9	7.0	7.1	7.1	7.2	7.2	7.3	7.3	7.4	7.4	7.5	7.5
5	.05	3.64	3.74	3.79	3.83	3.83	3.83	3.83	3.83	3.83	3.83	3.83	3.83	3.83	3.83
	.01	5.70	5.96	6.11	6.18	6.26	6.33	6.40	6.44	6.5	6.6	6.6	6.7	6.7	6.8
6	.05	3.46	3.58	3.54	3.68	3.68	3.68	3.68	3.68	3.68	3.68	3.68	3.68	3.68	3.68
	.01	5.24	5.51	5.65	5.73	5.81	5.88	5.95	6.00	6.0	6.1	6.2	6.2	6.3	6.3
7	.05	3.35	3.47	3.54	3.58	3.60	3.61	3.61	3.61	3.61	3.61	3.61	3.61	3.61	3.61
	.01	4.95	5.22	5.37	5.45	5.53	5.61	5.69	5.73	5.8	5.8	5.9	5.9	6.0	6.0
8	.05	3.26	3.39	3.47	3.52	3.55	3.56	3.56	3.56	3.56	3.56	3.56	3.56	3.56	3.56
	.01	4.74	5.00	5.14	5.23	5.32	5.40	5.47	5.51	5.5	5.6	5.7	5.7	5.8	5.8
9	.05	3.20	3.34	3.41	3.47	3.50	3.52	3.52	3.52	3.52	3.52	3.52	3.52	3.52	3.52
	.01	4.60	4.86	4.99	5.08	5.17	5.25	5.32	5.36	5.4	5.5	5.5	5.6	5.7	5.7
10	.05	3.15	3.30	3.37	3.43	3.46	3.47	3.47	3.47	3.47	3.47	3.47	3.47	3.47	3.48
	.01	4.48	4.73	4.88	4.96	5.06	5.13	5.20	5.24	5.28	5.36	5.42	5.48	5.54	5.55
11	.05	3.11	3.27	3.35	3.39	3.43	3.44	3.45	3.46	3.46	3.46	3.46	3.46	3.47	3.48
	.01	4.39	4.63	4.77	4.86	4.94	5.01	5.06	5.12	5.15	5.24	5.28	5.34	5.38	5.39
12	.05	3.08	3.23	3.33	3.36	3.40	3.42	3.44	3.44	3.46	3.46	3.46	3.46	3.47	3.48
	.01	4.32	4.55	4.68	4.76	4.84	4.92	4.96	5.02	5.07	5.13	5.17	5.22	5.24	5.26
13	.05	3.06	3.21	3.30	3.35	3.38	3.41	3.42	3.44	3.45	3.45	3.46	3.46	3.47	3.47
	.01	4.26	4.48	4.62	4.69	4.74	4.84	4.88	4.94	4.98	5.04	5.08	5.13	5.14	5.15
14	.05	3.03	3.18	3.27	3.33	3.37	3.39	3.41	3.42	3.44	3.45	3.46	3.46	3.47	3.47
	.01	4.21	4.42	4.55	4.63	4.70	4.78	4.83	4.87	4.91	4.96	5.00	5.04	5.06	5.07
15	.05	3.01	3.16	3.25	3.31	3.36	3.38	3.40	3.42	3.43	3.44	3.45	3.46	3.47	3.47
	.01	4.17	4.37	4.50	4.58	4.64	4.72	4.77	4.81	4.84	4.90	4.94	4.97	4.99	5.00

Table E (*continued*)

Error *df*	Level	Number of means in range being tested: 2	3	4	5	6	7	8	9	10	12	14	16	18	20
16	.05	3.00	3.15	3.23	3.30	3.34	3.37	3.39	3.41	3.43	3.44	3.45	3.46	3.47	3.47
	.01	4.13	4.34	4.45	4.54	4.60	4.67	4.72	4.76	4.79	4.84	4.88	4.91	4.93	4.94
17	.05	2.98	3.13	3.22	3.28	3.33	3.36	3.38	3.40	3.42	3.44	3.45	3.46	3.47	3.47
	.01	4.10	4.30	4.41	4.50	4.56	4.63	4.68	4.72	4.75	4.80	4.83	4.86	4.88	4.89
18	.05	2.97	3.12	3.21	3.27	3.32	3.35	3.37	3.39	3.41	3.43	3.45	3.46	3.47	3.47
	.01	4.07	4.27	4.38	4.46	4.53	4.59	4.64	4.68	4.71	4.76	4.79	4.82	4.84	4.85
19	.05	2.96	3.11	3.19	3.26	3.31	3.35	3.37	3.39	3.41	3.43	3.44	3.46	3.47	3.47
	.01	4.05	4.24	4.35	4.43	4.50	4.56	4.61	4.64	4.67	4.72	4.76	4.79	4.81	4.82
20	.05	2.95	3.10	3.18	3.25	3.30	3.34	3.36	3.38	3.40	3.43	3.44	3.46	3.46	3.47
	.01	4.02	4.22	4.33	4.40	4.47	4.53	4.58	4.61	4.65	4.69	4.73	4.76	4.78	4.79
22	.05	2.93	3.08	3.17	3.24	3.29	3.32	3.35	3.37	3.39	3.42	3.44	3.45	3.46	3.47
	.01	3.99	4.17	4.28	4.36	4.42	4.48	4.53	4.57	4.60	4.65	4.68	4.71	4.74	4.75
24	.05	2.92	3.07	3.15	3.22	3.28	3.31	3.34	3.37	3.38	3.41	3.44	3.45	3.46	3.47
	.01	3.96	4.14	4.24	4.33	4.39	4.44	4.49	4.53	4.57	4.62	4.64	4.67	4.70	4.72
26	.05	2.91	3.06	3.14	3.21	3.27	3.30	3.34	3.36	3.38	3.41	3.43	3.45	3.46	3.47
	.01	3.93	4.11	4.21	4.30	4.36	4.41	4.46	4.50	4.53	4.58	4.62	4.65	4.67	4.69
28	.05	2.90	3.04	3.13	3.20	3.26	3.30	3.33	3.35	3.37	3.40	3.43	3.45	3.46	3.47
	.01	3.91	4.08	4.18	4.28	4.34	4.39	4.43	4.47	4.51	4.56	4.60	4.62	4.65	4.67
30	.05	2.89	3.04	3.12	3.20	3.25	3.29	3.32	3.35	3.37	3.40	3.43	3.44	3.46	3.47
	.01	3.89	4.06	4.16	4.22	4.32	4.36	4.41	4.45	4.48	4.54	4.58	4.61	4.63	4.65
40	.05	2.86	3.01	3.10	3.17	3.22	3.27	3.30	3.33	3.35	3.39	3.42	3.44	3.46	3.47
	.01	3.82	3.99	4.10	4.17	4.24	4.30	4.34	4.37	4.41	4.46	4.51	4.54	4.57	4.53
60	.05	2.83	2.98	3.08	3.14	3.20	3.24	3.28	3.31	3.33	3.37	3.40	3.43	3.45	3.47
	.01	3.76	3.92	4.03	4.12	4.17	4.23	4.27	4.31	4.34	4.39	4.44	4.47	4.50	4.58
100	.05	2.80	2.95	3.05	3.12	3.18	3.22	3.26	3.29	3.32	3.36	3.40	3.42	3.45	3.47
	.01	3.71	3.86	3.98	4.06	4.11	4.17	4.21	4.25	4.29	4.35	4.38	4.42	4.45	4.48
∞	.05	2.77	2.92	3.02	3.09	3.15	3.19	3.23	3.26	3.29	3.34	3.38	3.41	3.44	3.47
	.01	3.64	3.80	3.90	3.98	4.04	4.09	4.14	4.17	4.20	4.26	4.31	4.34	4.38	4.41

SOURCE: Adapted and modified from D. B. Duncan, "Multiple Range and Multiple *F* tests," *BIOMETRICS* **11**: 1–42 (1955) with the permission of the author and editor.

INDEX

1 2 3 4 5 6 7 8 9 10